# CAN TECHNOLOGY INFLUENCE ECONOMY DEVELOPMENT

JOHN LOK

Made with ♥ on the Notion Press Platform
www.notionpress.com

# Contents

# Preface

Introduction

Our business society had developed long time from farming period to manufacturing period, then to service industry period, till to nowadays technology service and manufacturing period. It brings this question: Can technology or human behavior may influence economic development? Why does robotic can improve employee performance? How can employees behaviors influence organizational efficiencies when robotic participate to any organizations? How to evaluate or judge whether employee individual performance may satisfy organization needs? How to improve employee performance when AI participates to cooperate to labor to work together? What negative influences to be caused to the organization, if it neglected to raise employee performance or neglects AI participation to any organizations' tasks ?

In first chapter, I shall explains why organizations need strategy to be implemented as well as what business model means as well as why business organizations may experience different life cycle stages.

In second chapter,I shall explain that economic recession or boom how influences consumer behavior e.g. the business had been experiencing decline life cycle stage, such as COVID -19 disease occurrence. I shall explain how to apply business development strategy to raise the educational robotic manufacturer sale number.

In third chapter, I shall explain how to learn behavioral economy to solve social challenge as well as why some social challenges may influence customers number .

In fourth chapter, I shall explain what factors influence our tourism industry life cycle stage as well as whether how strateges may influence tourism industry develops.

In five chapter, i shall explain whether robots may help economic development.

In six chapter, I shall explain whether technology or human behavior whether may influence economic growth or recession.

# Prologue

Table of content

● Why do some social challenges may influence customers number ? p.46-60

Chapter 4

Can space tourism influence economic development

● New and old economic theories explain oil is not main factor to influence tourism income

● What are the characteristics of birth life cycle stage to tourism industry ? p.61-95

● What characteristics to space tourism growth stage?

Chapter 5

Robots How Help economic development

Artificial intelligence and customer service relationship

● How to apply robotic to raise customer service performance p.96-120

● How robotic improves client communication

Robotic how helps labours to avoid abnormal working hours p.121-138

● Why can robotic participation assist workers do not need to work overtime or need to work abnormal hours often?

● Robotic participation will bring these benefits

● How robotic reduces restaurant labours working hours

How robotic changes global future labour market

Applying management science equation to develop artificial intelligence to do management strategy

● How to apply management skills to let artificial intelligent robotic to learn

● Why does robotic need to learn customer satisfaction skill

● Worker lazy behavior applies to robotic

● How robotic and workers coordination in factory condition

● How and why robotic participation can bring personnel selection benefit

● Why can robotic bring talent working place benefit and reduces long time training and development cost

● How to implement the mind of strategy skills to robotic?

● How robotic can manufacturing and service job market?

● How artificial intelligence influence e-business workers market?

- How robotic becomes one kind of factor of production to the owning robotic task participation organizations?
- Why can robotic avoid productivity challenge and the firm does not need to employ extra worker ?

How robotic brings positive and negative social change

- Why does robotic seem to McDonaldization franchise sale method p.139-150
  - Why robotic seems to McDonaldization's operation?
  - Why does robotic seems to be common social commodity?
  - How can robotic bring global social stratification positive and negative change?

Robotic how influences global economic change

- Encouraging new economic development

competitive factor

- How much robotic productivity can raise competition to cause effects of limited competition disadvantage to our society.
- How and why robotic brings long time intellectual development to global factories ?
- How can robotic help manufacturers to raise productivities or improve efficiency?
- Why does production possibilities curve may explain65 why robotic participation may assist few workers to raise productivities in factories ?

Must Developed And Developing Countries
Need Artificial Intelligent To Replace Human Job

- How AI help developing countries to communication and agriculture and learning and medical delivery development 151-165

- Emergency Response to developing countries' earthquake natural damage sudden occurrence predicting
- Smart AI Agriculture
- Medicine Delivery to developing countries' patients urgent need
- Assistance to reduce teaching work workload or psychological pressure to teachers in developing countries' schools
- Why does smart phone help developing countries communication ?
- The Positive Impact of Mass Media in Developing Countries
- Why do developed countries need to develop AI
- AI may bring what benefits to developed countries

mean ?

The relationship between social change and human behavior

How human productive behavior may influence economic development

- New Zealand farmer individual wine productive behavior
- America high technological productive behavior
- China share market investing behavior

Why has any individual country have many people invest share behavior which can influence the country's macro consumption desire? p.204-220

Can technology influence human shopping behavioral change?

Why and how human behavior may influence the country's economic growth or recession?

Technology how impacts human behavior changing?

How and why employees behaviors may influence economy development?

Robots invention whether they can help organizations to raise efficiencies or inefficiencies?

Why social behavior may influence organizational strategy needs to be changed ?

How and why human behavior may influence economic growth or recession?

Reasons why human behavior may influence economic recession or growth ?

How employee behavior influences organizational development?

CHAPTER ONE

# Strategy function to organization

● Explaining what strategy means?

What does concept of corporate strategy mean ? Why does organization need corporate strategy ? The reasons may include : reducing cost, making reasonable or the most beneficial decisions or actions, earning above average returns etc. strategy may be a set of key decisions made to meet objectives. A strategy of a business organization is a compenhensive matter plan stating how the organization will achieve its mission and objectives.

A successful strategy may have these four perspectives, a plan, how do I get these; a pattern , in consistent actions over time; a position, it reflect markets, a ploy is a maneuver instead to outwit a competitor, a perspective is a vision, direction, a view of what the company or organization is to become. For minimizes or competitive disadvantage strategy example, company realizes merging with companies advantage . Although, it may not make its market leader , but it may venture into retailing will help it increase profit.

Strategy also may provide a clear understanding of purpose, objectives and standards performance to employees at all level in all functional areas. Usually, every firm competing in an industry as a strategy, because strategy refers how a given objectives will be achieved. For example, computer industry uses a differentation competitive emphasizes innovative product with creative design. For example, coeporate strategy, Coco Cola Inc. has followed the growth strategy by acquition. It has acquired local bottling units to become as the market leader. For function strategy example, pocter and Gamber spends huge amounts on advertising to create customer demand . They aim to maximize resource productivity. It is concerned with developing a distinctive competence to provide the firm with a competitive advantage. Thus, strategy may have different functions. It depends on whether the organization needs what strategy to achieve its objectives or aims.

● Why does organization need strategist

However, any organization needs one or more strategist (s) . strengths are individuals or groups who are primarily involved in the formulation, implementation , and evaluation of strategy. In a limited sense, all managers are strategists. Strategists may include: Consulants, entreprensurs, boards of directors, chief executive officer, senior management, corporate planning staff, strategic business unit level executives, middle level managers, executive assistant titles in any organizations.

Organizations needs outsourced consultants service because many organizations do not have a corporate planning department, owing to small size . Thus, outsource consultancy firm can provide this kind service to them.

Entrepreneurs are promoters who conceive idea of starting a business for getting maximum returns on investment. They are awaiting for an environment change and for an opportunity in the best interest, for example, a biotechnology firm's managing director needs to implement policy formulation in research and development department.

Board of directors are professionals elected may by the shareholders of the company as per rules and regulation of the company act. They are responsible for the general administration of the organization. They are supposed to guide the top management .

CEO is the top man, next to the directors of the board the occupies the most sensitive post, being held responsible for all aspects to strategic management right from formulation to evaluation of strategy.

Senior management from the chief executive to the level of functional or profit centre heads. They are involved in various aspect of strategic management .

Strateic business unit is diivided into different independent units and allowed to form own respective strategies. Middle level managers are operational planners, for departmental plans, as implementers of the decisions as well as executive assistant is a person who assists the chief executive in the performance of his duties in various ways, e.g. data collection and analysis, suggesting alternative, where decisions are required. preparing briefs of various proposals, projects and reports, helping in public relation. All of above positons may be any organizations' role in strategic plan.

● Why do SWOT ( strengths, weaknesses, opportunitites and threats) analysis can remain a major strategic tool to any organizations?

It is one straight formed methodology for making a structured analysis of strengths and weaknesses into core competences and core problems by using the core-competence tree and the current reality true. The core competences and core problems are then linked into a plan of action aimed at preserving the organization's core competence. It supposes that any organizations ought have internal strengths and weaknesses both. So, if the organization has strengths , it also ought have weaknesses. Any organization, itself ought have ability to control or avoid or threaten its any weaknesses cause as well as finds any method to arise strengths to bring itself competitive ability. Otherwise, due to external environment factor, it can not control. So, it supposes any organization can not control any opportunities ot threats when they will occur or encounter to influence weaknesses, eliminate all weaknesses that do not satisfy the following criteria.

The weaknesses that do not satisfy the following criteria: The weaknesses must exist over a period of time can not be a one-time phenomenon, the weakness must be expressed in undesirable terms, the weakness must be under the firm's control or influence . So, any firm hopes to eliminate weakness, it depends on how it causes significant damage to the company. For example, when one firm discovers that the project ought may be finished within five months. But, after four months, it discovered that this project can not be finished, if it hopes staffs can cooperate to finish this project before five months, it needs to find whether what its main weaknesses are influenced this project will delay, e.g. lack of innovation, lack of growth, insufficient attractive profits to excite staffs to work. Hence , SWOT is a strategic management tool, it consists of the analysis, decisions and actions, an organization undertakes in order to create competitive advantages. However, the next phases of the strategic management process is external and internal organization's strengths and weaknesses analysis, by conducting an external analysis , an organization also needs to identify the critical threats and opportunities in its competitive environment. It also needs to examine who external competitive environment influences its business develops in long term.

In fact, any organizations need to make the most reasonable strategic choice with vision, mission, objectives and the external and internal analysis of its external environment influence. Hence, the strategic management process may include this steps:

From vision to mission to objectives to ( SWOT analysis, external analysis and internal analysis both ) to strategic choice ( the most reasonable choice) to strategy implementation to achive competitive advantage . So, any organizations must need to spend long time to gather data to analysis whether which it has actual internal strengths and weaknesses as well as what the present external envioronment brings opportunities and threats to influence its business development, if it hopes to implement effective strategic plan. So, it seems that SWOT ought be one step to any organization's strategic plan management process. Thus, managers have responsibilities to help their organizations to try to " fit" the analysis of externalities and internalities, to balance the organization's strengths and weaknesses as well as environmental opportunities and threats . For one car sale manager hopes to find methods to solve its car low sale problem. In the SWOT analysis, it may have these questions: Why does the performance of the car firms in the same motoe sale service industry, operating under the same competitive environment? Which tangible resources of the high performance motor sale service firm provide sources of competitive advantage and subsequent superior motor sale service firm performance? How do the identifical tangible resources actually create value for a motor sale service firm in the motore service industry and provide the motor sale service firm with source of sustainable competitive advantage?

Thus, assumption of questions are needed in order to help organizations to attempt to seek the main factors influence their short or long objectives can not achieve in the SWOT analysis process. Then, they can evaluate the different factors to make the reasonable analysis to decide whether which is the main factor to influence their objectives can not achieve satisfactory . Then, they can revise their errors as well as find the most reasonable or the most right solutions methods to achieve their objectives more successfully. Because some factors influence the business actions its objectives succussfully. They may include: poor organizational behavior, e.g. worse staff performance, working attitude, lazy , they do not enjoy or feel bore to so their work, they feel salaries are not reasonable; poor strategy, e.g. the business ought not expand more branches at this moment rapidly, the business ought chose partnership , it is more suitable to compare sole trader formation, the organization ought advertise to raise its brand awareness to let public to acknowledge etc. wrong strategy implementation. So, SWOT stragegy role may also help any organizatons to revise whether they have errors in order to find the most reasonable factor to cause their poor performance or low profit etc. effects.

- What is strategy and strategic management to future managers in organizations?

Are they understood and recognized? However, I believe that the development of organizational strategy depends on understanding the perceptions of their managers on what strategy and strategic management actually is. The identifications of perceptions of future maangers will need have some insights , opinions and knowledge on the organization's this matter reflect the efficiency and effectiveness of the strategy related learning proces in themsleves organizations. So, I believe that, the manager needs have enough knowledge about how to manage the kind of business if he/she hopes to become the organization's proficient leader, for example, one supermarket business CEO ought own part supermarket operation experience, when he/she has practised supermarket operation experience to know how to manager teams cooperation efficiently, e.g. cashiers, food promoters, food warehouse delivers, grocery shelf putters, fresh fish and fruit pick up keepers, family daily e.g. tooth paste, bath daily products, washing cloth products shelf putters staffs. Then, the supermarket store manager ought have excellent managing ability to manage the supermarket different sale teams to cooperate efficiently. So, owing the kind business managing experience to the manager will be one main factor to influence the business to grow or expand more successfully. Thus, how management develops strategies to guide how an organization conducts its business and how it will achieve its target objectives . The manager himself/herself managing ability and the organization's target objectives can be achieved, they will have close relationship . It is management's responsibility to adjust negative conditions by undertaking strategies defense and managerial approaches that can overcome adversity. However, the essence of the good strategy-making is to build a positive strong and flexible enough to provide successful performance despite unforeseeable and unexpected external factors.

Thus, the five tasks of strategic management may include as below: First step, developing a vision and a mission. It means that any firm ask is " What is our business and what will it be ? What is our business and what will it be ? " Managers need to develop the next five to ten years a clear mission to his organization needs to achieve. A clear mission can establish the organization's future effects and outlines " Who we are, What we need to do and where we are going ? "

Next step. setting objectives, or mission statements can achieve performance targets more easily . Objectives serve as for tracking an organization's performance and progress . A desired performance can pushes an organization to be more incentive, how to improve its financial performance, and its business position. Objectives may have short, middle and long time. Objectives may have two kinds. One is financial objectives, e.g. measures as earning's growth, return on investment and cash flow. The another is strategic objectives, it provides consistent direction in strengthening a company's overall business positon. They relate more directly to a company's overall competitive situation, such as growing faster than the industry's average and making gains in market share.

Then, crafting a strategy step, it is a SWOT analysis to find how to achieve organizational mission. Thus, strategy will be the most important part in order to let the organization to achieve its short, middle and long objectives more easily. For motor industry example, global competitor is competitive environment to motor manufacture industry, if the motor firm can not innovate its any kinds of motors, then it can not attract car buyers to choose its cars to drive. So, in vehicle manufacturing strategy aspect, if the vehicle firm hopes to raise its any kinds of car sale number. How to innovate to manufacturer " new design cars" which may be one important factor to influence any one motor firm sale growth in success. Then, car industry strategy ought concentrate on how to improve the car design to be more attractive. It is " the growth of motor design techniques concep strategy to any nowadays car manufacturers ". It seems that is SWOT analysis to car industry internal strengths and weaknesses analysis have more influential to compare external environment opportunities and threats to any one car manufacturer seller ought not only consider how to train car salespeople sale skill. They ought consider how to provide the useful opinions from car customers' design feeling to car manufacturers, in order to assist them to attempt to design any the most attractive car design to satisfy car buyers' needs in this global car competitive market. So, car design will be one important factor to influence car sale growth. Strategy ought focus on " how to innovate car design" to satisfy car buyers' car design pursue.

On conclusion, any kinds of businesses must have themselves characteristics or features. So , management ought need to consider how themselves business features or characteristics to decide the most reasonable ot the most right strategy implementation in order to achieve their objectives or missions more easily. So, lacking any strategic organizations ought be difficulty to achieve their missions or objectives to compare owning any strategic organizations in nowadays business environment.

● What does business development strategy ?

An effecting business development strategy ought have these five steps: The first step is market analysis. Who are your clients , knowledge of your market? Second step is how to adopt for each penetration, your business needs to learn how to adopt for each group of clients, your first need to review your own capacbility. It is important that you are realistic and honest with yourselves over where clients truly sit, learn how to classify your clients into similar groups relative is the scale of the opportunity. Third step learns how to review your performance , market matrix to plot your results to help you determine your market penerstion. In addition, it will help you then discuss and consider various strategies for growth. By potting your clients you will get a sense of where your strengths and weaknesses are against the opportunity that total market offer.Fourth step learns how to consider alternative growth strategies on the market matrix. The final step , you need to consider these questions in order to decide whic is the most effective strategy for your business. For example, which model is the most ( least effective? Why? which model work best for line managers, HR are finance, why? How might we most effectively progress from one model to the most reasonable questions? ) Then, you will need to decide how to launch new services, new products, opening new markets, how accessing new geographic territories.

● What is business model?

It is logic and provides date and other service evidences that demonstrates how a business creates and delivers values to customer. It also help how to predict revenues, costs, and profit with the business enterprise delivering that value. How does on build a competitive advantages and a super normal profit ? How the enterprise creates and delivers value to customer, and receive payments to profit easily .

In essence, a business model is a conceptual, rather than financial model of a business. An effective business model may help you to decide how to create value for customers, receive payment to profit more easily. For example, driving factors include knowledge economy, the growth of the internet and ecommerce, the outsourcing and offshoring of many busness activities, and restructuring of the financial services industry, i.e. the enterprise simply need to learn packed its technology and intellectural property into a product which it sold, either as a discrect item or as a bundled package.

The existence of electronic computers that allow low cost financial statement modeling has facilitates of assumptions about future revenues and costs. Also, the concept of a business model has no established theoetical in economics or in business studies. Economic theory assumes that trades take place around tangible products : intangibles are the best. For example, inventions are often assumed to create value naturalty and enjoying protection of patients, firms can capture value by selling patients to market, i.e. the publisher sells the another's books, the books can help it to earn high level of royalty income. In economic theory explains the publishers can create intengible value, e.g. royalty as well as tangible value, e.g. selling books.

However, business mides are necessary features of market economies, it is consumer choice, transaction cost, amongst consumers and producers and competition. It meets invention and consumer wants of new product value need and the opportunity to satisfy their needs. So, good designs are likely to be highly siutational, and the design process is likely to involve processes. New business models can both faciliate and represent innovation. For example, in te sport apparel business, sponsorship is a key component of today's business models, Nike, Reebook, Adidas and other sponsor football and rugby clubs and teams as well as royalties from sale sport related products, e.g. sport shoe, spot cloth. Moreover, business models must be over times as changing markets , technologies and legal structures to adopt the kind of business market change.

A business model achieve the logic the useful and reasonable data and evidence thst support a value decision for the customer. In practice, successful business models very often become to some degree, "shared" by multiple competitors in possible . Strategy analysis is this an essential step in designing a competitive business model, i.e. low cost strategy for newspaper advertising ( including classifieds) helps cost of generative content is easy to replicate and of many different geographically separate newspaper market in the world. So, when a country's newspaper publisher may have a differentiated and hard to initate low cost advertising strategy, but it can achieve the same time effective and efficient. Its business model will be successful in newspaper publishing industry. Hence, it seems that

business model will be any businesses‘ essential part in their growth strategies. If the business has none a successful business model, it will not have successful growth strategy consequently.
Growth strategy is different to business development strategy, why ? Growth strategy will mark afresh start, by having all economic actors in the private sector activity and dynamically undertake efforts to promote growth with a determination to take on challenges, when the business feels that it is the right time to grow its business. Otherwise, business development strategy is not the businessmen's feeling whether it is the right time to develop its business . It is essential part to any business expects to start.

- What does business climate development strategy ?

What are the different between business growth and business development and business climate development strategy ? In fact, business growth strategy refers any businesses start up in beginning from earlier stage to nature stage of life cycle. Otherwise , business development strategy must not start up from beginning. It is common on the middle stage, the business hopes to develop its market share or new market to be more. So, it needs to find whether what its SWOT in order to develop its new niche market more successfully.
Hence, it brings this question? Business climate development strategy is on the beginning or middle or mature stage in life cycle? I shall explain as below:
A business climate development strategy means that it is one targeted policy tools appear to favour medium-sized , well established industrial enterprises over younger, small enterprises with high-growth potential operating in the services. Hence, any governments may attempt to follow the business environment to implement any methods to help small size businesses to grow up easily in the beginning stage. In general, implement to business elimate development strategy challenges may include: Lacking of coordination means that there is a disconnect between business and innovation support policies on one hand of investment localization on the other. However, to solve this challenge, governments may encourage these organizations to participate , such as technical centres, laboratories and training facilities can act as catalysts for industry, sector aggregation and support the establishment and specialized investment zones. Hence, the difference between business climate development strategy and the other both strategies. Business climate development strategy is any countries' governments attempt to follow the business environment at the moment to

find the most effective methods to help any small size businesses to grow up more easily in the beginning stage.

How to implement business climate development strategy more successful? The key recommendations may include: To emergy from the workshop was further strengthened in an effort to close the policy gap and create synergies between programmes. Building on the experience of regional investment centres, the context points would identify an small middle size's (SMS) organization, and help enterpreneurs to establish a network of SMS bisiness centres across the country. So, they could not as single -window contact points, storing and channelling information abour all the government's SMS business programmes. However, a successful business climate development strategy should build on the analyzed risks and rewards of informal business operations and aim at modifying the behavior of economic agents ( enterprises, employees and customers) through a combination of incentives and penalties.

Any governments may attempt to develop a number of measures to promote and suport innovation and upgrade technology in the private enterprise sector. In order to enhance business climate development strategy to implement in success. They may include: establish a system of communication and cooperation between the institutions and private sector organizations operating in the area of technological upgrading, innovation, financial, technical standards, public education and training . It would be useful to conduct an evaluation of the enterprise Europe network's impact in order to learn from the lessons of the country's more successful sector-specific centres. It is the national strategy for innovation, technology upgrading and investment encouragement. Finally, a critical element in an effective innovation strategy is the establishment of links between support services and programmes and access to funding to support any business founder hopes to develop himself business in success in order to adopt business environment climate change more easily.

However, in general, small or medium size business will encounter these challenges when they hope to adopt the business environment climate change to grow their businesses more easily. Their challenges may include: The lack of economies of scale, which limits their ability to invest in fixed capital and technological development, proportionally higher costs,, which increases the impact of the legislative and regulatory framework, information lacking which limit access to external financing, limited resources for internal training and human capital developments. In general, any governments need to help the new business to solve these challenges in order to develop business climate development strategy more easily. They may include: innovation technology centres and networks as well as financial support for innovative SMES. Hence, business climate connection with larger foreign enterprises through active government support, the policy objective is to enhance SME access to international markets, skills development, finance and technology to any businesses' beginning stage in order they can grow up their businesses to nature stage in their business life cycle successfully.

Hence, whether which firms hace real need to get government's business climate development strategy assistance. I believe that the government needs to assume the firm has these challenges in order to ensure that the firm can achieve the requirement to get this business climate development strategy assistance. They are needed to assume on basic these factors: A company must grow and pass through all stages of development or die in attempt, second the models fail to capture the important early stages in a company's origin and growth. Third, instead of annual sales , although some mention number of employees whether it is more or less factor, government can not ignore other factors to decide whether the firm can be accepted to implement climate business development strategy assistance, such as value added, number of locations, complexity of product line and ratio of change in products or production technology etc. factors to decide whether the firm is suitable to be accepted to get business climate development strategy assistance from the government in order to avoid waste time and resource and money to desing any kind of business climate development strategy to assist the firm to develop in the beginning.

● How to implement successful organizational downsizing strategy?

What are the effects of downsizing on organizational performance? What is the most right time to downsizing to the organization ? When one organization has grown to the bigger size , e.g. more revenues, this year than last, a larger workforce, greatest market share, downsizing strategy ? If an organization did not grow, it was viewed as stagnating and upproductive in the non-growth life cycles stage, it implements downsizing strategy to influence its performance to be worse.

Organizational downsizing strategy is one part of the management of an organization and designed to improve organizational efficiency, productivity and/or competitiveness. Downsizing means to reduce organizational size, e.g. staffs number reduces expenditure reduces, cost reduces . Downsizing is an intentional set of activities, it differentiates from loss of market share, loss of revenues or unwritting loss of human resources organizational decline. Also, downsizing usually reductions in personel, such as transfers outplacement, retirement incentives reduction, byout packages, layoffers. This reductions in personnel may occur in one part of an organization, but not in other parts, e.g. in the production function, or not in the engineering function. Finally, downsizing may effect work processes, e.g. fewer employees are left to do the same amount of work, and this has an impact on what work gets done and how it gets done. However, instead of downsizing of reduction employees number aspect, it may also occur on other accepts, such as selling off, transferring out, merging businesses or altering the industry structure. It aims to improve organizational performance. Labeled workforce reduction strategies, focused mainly on eliminating headcount or reducing the number of employees in the workforce. It aims to early retirements, transfers and outplacement, by-out packages. This kind downsizing strategy whether it can bring performance improving benefit to organization or not in long term? Less employees work whether it will still improve performance, although the organization can reduce salary expenditure . Otherwise, if the organization does not reduce staffs number, it chooses to workforce reduction, work redesign and systemic strategies , whether it will be better than staffs number reduction strategy ?

What are critical success factors influence any organizations' strategic downsizing success implement ? Addresses the rationale utilized by firms to downsize, the expected outcomes in terms of economic and human consequences, and specific strategy. Also , downsizing tactics, human resources as assets to cost planning, participation, leadership, communications and support to victims . survivors are examined to any attempt implementing downsizing organizations.

In past organizations, when many blue-collor workers are also to trade off wage freezes for jobs security. White-dollars workers in the lower ranks of white-collar workers are often dismissed by downsizing, due to firms are increasingly forced to cut costs, restructure, and reduce their labor force. Instead of western countries firms are popular to accept downsizing . Downsizing has even become common in industrialized countries, such as Japan and Sweden, restructing in the 1990s led to employment reductions in industry and thus, increases in the levels of unemployment. Hence, downsizing may cause low ranks of white-collar workers feel job security

lose in any time when they are working in any organizations, because white-collar workers are different to blue-collar workers have unions protection.
However, there are three perspectives from when downsizing can be reviewed : The industry level, the organization level and the individual level in terms of industry or global perspective, it may include , mergers, acquisitions, joint ventures, the organizational and strategy level may include how to implement downsizing and the expected bebefits of downsizing on the firm's performance, efficiency , and at the individual psychological level, it includes employee himself/herself stress, negative emotion feeling , due to he/she is dismissed. Hence, any organizations need to considerate how downsizing brings negative emotion to influence every dismissed employees. Because , their leave which will influence the continue working employee's emotion feel fear to be dismissed in next. If the present employees often feel stress to work, then their performance and efficiency will be influenced to worse. So, any organizations can not neglect to care the current working employees individual emotin in order to avoid low efficiency and poor performance to their organizations. Becaus every employee will have possible to be unreasonable dismiss, due to downsizing organizational influence.

On possible reason for this occurrence, is that firms poorly planned or carried out earlier downsizing projects and hence must remedy past facilities. So, one planned downsizing strategy will avoid negative emotion to influence current employees' works. But, factory workers will have possible to encounter dismiss , due to technological improvements, e.g. robotics can reduce to have additional workers rather than replacing the existing employees. So, when the factory begins to apply robotics , then employees number will be reduced. It is technological manufactuer causes downsizing to factory workers reason. It is due to raise productive efficiency factor , more than reducing cost reason to cause downsizing need to any organizations.

The term downsizing was first used reforcing to strategies to reduce personnal. However, it was become more and more relevant , its scope has been expanded and noe refers to a wide range of management measures towards better adopting on organizations to its environment ( Gandolfi & Hanson, 2011).

reference

Gandolfi , F. & Hanson, M. (2011). Causes and consequences of downsizing : towards an integrative framwork, Journal of management & organization , 17(4), 498-521.

In general, it is needed to implement strategies, due to the organization feels that without achieving the required organizational changes, this failing toobtain the desired results ( Magan & Cespeses, 2012).

The downsizing methods may include: retrenchment specialized production, concentrating activities until economies of scale have been achieved. Downscaling strategy is toward again reducing in a smaller differentiation of activities in the value chain , it aims to keep the organization to reduce complexity in the organization. For some organizations had begun to implement robotic factory, because they expect that manufacturing robotics can help they to specialize production, raise productive efficiency. Hence, they only concentrate on keeping the proficient workers, they can cooperate to robotics to work in order to raise more products number efficietntly every day. So, the low skillful workers will be dismissed and they will re-employ the owning control manuacturing robotics skillful new workers to replace them. So, future manufacturing robotic manufacture plants causes downsize, it will be one good example for specialized production, concentating activities until economies of sale reasons to cause downsizing factory workers need to the owning manufacture robotics plant organizations. So, downsizing has an impact on raising productivities on specialized production and concentrating activities until economies of scale aim more than reducing cost to the owning manufacturing robotic factories organizations.

Thus, downsizing activities aim to improve organizational efficiency, productivity and/or competitiveness that affect the size of the firm's workforce, costs and the work processes. Downsizing may include: building-down , de-hiring, de-recruitment, reduction in force, re-sizing and right -sizing. So, in macroeconomic factors view, global competiton , technological innovations ,a change in business strategy retains competitive advantages may cause why some organizations decide to implement downsizing strategy.

However, one successful downsizing strategy implements to any organizations, organizations can not only consider themselves benefits, they also need to consider the psychological contract between employer and employee as new mutual expectations on workplace environment. Frequently described as re-organization, restructing, downsizing or real sizing, the human resource effects of these changes have often been very destructive to individual lives, employment relationships and organizational efficiency. If mployees recognized that their company was creative and consistent action to pressure their employment ( security, trust could be reestablished and the success of the adoptive strategies. They can feel that their organizations decide to achieve the downsizing strategy is very reasonable more than unreasonable strategy in the right time. For example, when the organization decides to implement downsizing srategy before, they may enquire to their staffs opinions and acknowledge what their emotions, e.g. information on when change provided, staff views on the change are sought and are acted upon, staff have the opportunity to voice disagreement, support from manager during the change, to let they feel that change process seen as fair and equitable to let staff feel job security during the change process, and staff are trained to meet new job roles. So, all these factors may influence present employees have confidence to continue to work in your organizations after downsizing strategy is implemented. So, any organizations can not neglect to consider their present employees' feeling or emotion in order to avoid many staffs decide to leave their organizations after downsizing strategy in implemenation later.

reference
Magan , A. & Cespeses, J. (2012). Why are Spanish companies implementing downsizing. Review of business 32(2), 5-22.

- Business growth strategy

What factors cam affect the performance and growth to small businesses? Why and how obstacles are problematic for growth? How these differ between micro, small and medium sized businesses? How the obstacles are shaped?

Any businessmen had a substative growth ambition, but it can not represent that growth ambitin must sicceed to grow up their businesses. However, they must need to solve challenges when they expect grow their businesses successfully. The unpredicted external environment , include market changing and the vision of the owner and their attitudes towards growth will influence whether their businesses can grow up in success.

Hence, if the new business can keep negative growth, it ought may suceed. Some strategies , business owners need to consider that these factors will obstacle to their growth, such as during a recession, their businesses ought be difficult to growth, strategic planning is only useful when the business has a definit objective in mind, investment in research and development is too risky, expensive and difficult for a small business, there is no way,we can improve cashflow situation, factoring is only useful, if you ave in trouble, employees do not want formal, pay-related incentives and they are no use in helping business grow, we do not need to engage the staff in a structured, involving way, we can not get recruits to fit our needs, our business don't need to restructure our management as our business grow.

● How whether what obstacles can affect small business growth?

I assume small businesses have general staffs number with 50 or less than 50 staffs . Also, growth ambition was higher among younger business owners. The business owner personal poor time management factor may be one obstacles, for example, a lack of management time was rated to be the most difficult obstacle for potential exporters, which is something of an obstacles for the significant exporters, little knowledge of how to export and difficulty in finding customers also attract higher ratings from potential exporters, e.g. the fear of payment problems, the cost of exporting and being too small to export are rated as being less significant obstacles by potential exporters and their perceptions are not to distant from the significant exporters. So, lack of management time and little knowledge of how to export may be the business ower's significant obstacles.

However, many small businesses are facing values number reducing challenges. How adoption to improve sale performance? Sale improvement strategy may include: appreciated new customers, more advertising, devised a new marketing strategy, dedicated sales/ marketing manager, undertaken training in marketing sales.

Overall, making this transition from being a micro business to a small business clearly requires a greater confidence in dealing with such matters, though, undertaking activity more frequently or simply the earger scale of the business necessitating increased familiarity and competence. Also, employing a professional manager can be seen as generally enabling a company to improve.

The aim of focus groups was to explore a company to improve owners' views on growth, including how they conceptialize growth, perceived barriers,the consequences of growth and personal cirsumstances and evidence of mindsets among owners which may restrict their potential business growth. In general, family owning-business strategies may include: maintain quality and higher prices, rather than lower prices raising the value added of products and shifting into a less marketplace and the emphasis towards areas where they sold direct to end clients, rather than acting as subsontractors, whose margins were being squeezed, up-selling in terms of volume or value to existing customers , e.g. a catering establishment noted then they tried to encourage customers to return, or an accountancy practice and a range of extra services.

Many owners did also acknowledge that the would likely more if they were more actively intending to grow business or if they saw evidence of potentially opportunities. Several those lacking a current plan were aware of a growing need to develop both a more strategies outlook and more formal systems, because of a general neglect of strategic thought, with a number noting that years of unplanned growth has left them realizing that the development of the busines, needed to catch up with the situation that they had found themselves in.

- What can impact on growth strategies on business?

Overall, making this transition from being a micro business to a small business clearly requires a greater confidence in dealing with such matters, through undertaking activity more frequently or simply the earger scale of the business necessitating measured familiarity and competence. Also, employing a professional manager can be seens as generally enabling a company to improve.

The aim of the focus groups was to explore in depth business owners' views on growth, including how they conceptualize growth, perceived barriers, the consequences of growth and personal circumstances and evidence of mindsets among owners which may restrict their potential business growth. In general, family ownin, business strategies may include: maintain quality and higher prices, rather than lower pruces raising the value added of products and shifting into a less marketplace and the emphasis towards areas where they sold direct to end clients, rather than acting as subcontractors, whose margins were bring squeezed, up-selling in terms of volume or value to existing customers , e.g. a catering establishment noted that they tried to encourage customers to return, or an accountancy practice and a range of extra services.

Manyowners did also acknowledge that they would likely plan more if they were more actively intending to grow business, or if they saw evidence of potentially opportunites. Several of those lacking a current plan were aware of a growing need to develop both a more strategic outlook and more formal systems, because of a general neglect of strategic thought, with a number noting that years of unplanned growth had left than realiaing that the development of the business needed to catch up with the situation that they had found themselves in.

- What can impact on growth strategies on business ?

Growth is important and key on survival of any business profit venture. Formulating and implementing effective growth strategies may enhance business pforit to any dynamic organization . Developing growth strategies to attract human resources, including increase in the sales volume per annum, an increase in the production capacity, increase in employment, increase in production volume and increase in the all of material, increase energy and power, these factors may influence the business's strategy is implemented effectively in order to achieve growth aim.

Growth strategies that a business enterprise may wish to adopt include: understanding customer expectation, service, positioning, market segementation, setting measuring market standards, relationship marketing, human resource strategy and successful communication strategy these factors may influence business growth success. When one organization ensures that it can achieve growth, it may evaluate to measure its overall performance by these several aspects, they may include sales, assets base, employee retention goodwill and increase business profits that drive investment and economic development. Business growth may introduce new products and services, or adding new features to existing products. Growth could also mean expansion of an organization in order to buy new assets develop new products or service to enhance new investments in the economy. I shall refer some growth strategies as below:

Market penetration strategy focuses on expanding sales of a company's existing products or services in an existing market. It may attract new customer for the products and increase the usage or purchase rate of existing customers , it is often achieved by increasing activities through more intensive distribution and competitive pricing promotion.

Market expansion or market development means to move it into a completely new market. This strategy is about existing product which are offered in a new market when a region business wants to expand, or when new markets are opening up, or new use is found for the existing product.

Product expansion or product development strategy means introducing a new idea into a company's existing market. It offers new products to an existing market. It tried to grow by developing improved products for the present market.

Diversification means companies with sell new products or new market. It is very risky strategy . It needs to research market to determine if consumers in the new market will potentiallu like the new products. Acquisition means the purchase of one company by another company. It may be private or public.

The new growth strategy can be used as an alternative channel. It involves pursuring cutomers in different ways for instance selling a company's products or services online . Through the use of the internet a customer can access products or services of a particular company in a new (alternative). It goes beyond envisioning a long-term success. It has to be follow these steps: establishing a value for the company, identifying an ideal customer , who is loyal to the company, defining a company's key indicators, verifying revenue streams for cost reduction, seeking competition in the external environment, focusing on company's strength ( internal), investing in talents ( effective human resources who are creative and innovation). So, managers must achieve on growth strategy to enable stakeholders not only to plan, but also to track organic growth in their revenue and allow effective and efficient allocation of resources toward a more centered effort to adapt to frequent changes in the company and the industry occasioned by technology and the differences in competition.

Hence, of growth strategies are effectively formulated and implemented according to indicators and plan, it will lead to increase a profit in that organization . Growth strategies are often called the master business strategies, they provide the basic direction for strategies action. They are the coodinated and efforts indirected towards achieving long term business objectives and profit. Growth strategies have played central roles in the expansion and profit. They have enabled organizations to increase market shares, develop new markets, and develop new products and services, so business profit will continue to increase economic development.

- Why and how can organizational life cycle models influence organizational performance?

What does organizational life cycle mean? Must any organizations have life cycle? What do the influences when the organization reachs the organizational life cycle stage? Can the organization implement any useful strategy , if had ability to known whether what stage is its organizational life cycle? I shall explain as below:

In general, organizational life cycle has three stages: Birth, young, and maturity or decline. The related goals of profit, growth and survival seem to have overall goal structures of most organizations. IN general, most organizations will experience all three stages. However, not all organizations pass through all three stages. In fact, only about one-half of all new busness, organizations survive longer than one and half years. Relatively few for profit or not -for-profit organizations survive long time to experience all three stages. I assume that profits growth is one main factor to influence any organizations whether it can experience long whole three stages in their organizational life cycle.

What is the three stages chacracteristics of organizational life cycle? In birth stage, a merger or a point venture may occasionally lead to the creation of a new organization. A organization may be either a single person expands or an entrepreneur cooperates people to help promote a new idea, product or service. The motive in both cases is usually the desire for profit. In youth stage, when professional management is taken over by a family with a controlling interest, the organization's primary goal often changing from profit to growth. The new management team wants to demonstrate its competence and growth is the most obvious aim. For example, a manager of a large organization must consider how much company's return an investment in the organization's growth stage in this new growth stage. It has these characteristics: goals become less specific, less measurable, increasing emphasis on marketing, hoping for the increases sales that will justify the expansion of plant anf acquisition of new, more effecticient tools and equipment. Finally in the maturity / decline stage, as an organization matures and starts to decline, a desire to survive which will be the organization's goal in this stage. Why does organization may experience this stage, the organization can evaluate whether it is experiencing this stage, depends on these factors, e.g. when organization increases large , its technology is complex, its structure is bureaucratic, it is financially oriented, it is greatly affected by market and social forces and it is so complex when it grows up its organization, it will increases more new departments and it will employ many extra staffs and it will create many new positions. Hence, it explains why the organization can survive long time, and it can experience all three stages, it must be more succefful to compare the another organization can not experience all three stages , because it is common that when the organization can experience all three stages. It must survive long time. Also, it means that it can earn profit growth . Otherwise, when the organization can only experience birth stage or youth stage , then it can not survive long time and without profit growth in possible.

Hence, when one organization can experience all three stages. It may be one successful profit growth organization, e.g. although one sale trade earn less profit, and its organization size is small, but the jobb trader's business can survive a long time. So, he/she sole trading organization may experience birth, young, and mature or decline stages. So, organizational life cycle model can be applied to large or middle or small size organization, even sale trader, partnership organization to help them to evaluate whether their businesses are experiencing which stage in order to implement the most suitable or reasonable strategies to help them to solve present challenges more easily.

● What is the essential elements of life cycle model to assist business development ?

Although, when one organization feels it encoounters the decline stage, it means that its organization has possible that it can not continue survive. But if it has good strategy to help itself to solve present challenges. It has chance to renew or continue to develop its business functions to be better. Then, it has chance to continue survive. Usually, when the business is experiencing decline stage, it ought decide to change its business direction in order to continue survue . For example, raising its capabilities of organizational learning and innovations, creating new profitable and vision into the renew survival stage, and increasing its competitiveness in cost. However, the decline stage is characteristics by deterriorating profits and a loss of market share. The renewing firms have a rebuild their learning and innovative capabilities and shape a new profit direction for business. The contribution links interactions of development more effective business functions to provide a tool to help the experiencing decline stage organizations to learn how to implement new strategies to solve their present challenges in order to continue survive.

Hence, what is the essential elements to help the experiencing decline stage organizations to have possible to continut survive. I shall explain as below: In old economic society, manufacturing firms whose main driver to standardize production, products and busines processes. By constrast, the new economic society, we are experiencing information business, utilize information to differentiate, personalize and dispatch over networks at an rapid speed, small as e-commerce. It is obvious tht old economy's traditional shop business model is not popular to be accepted by consumers. Consumer shopping behaviors have been changed to choose online shopping to replace visiting shops shopping behavior. So, it seems that why some businesses will experience decline stage within one year. It is possible that their traditional business visiting step shopping method is not popular to be accept to themselves businesses, it is right time.

They need to design website store to let customers can choose online shopping when they visit themselves website stores from internet. So, e-commerce can influence some businesses to experience the decline stage in short time rapidly. It may be one factor to influence any organizational life cycle stge to be shorten in short time. It is one technological innovation element to shorten any organization's life cycle in short time. So, any businessmen ought not neglect technological innovation new influence their businesses development as well as they need to continue pursue their new sale method direction, e.g. payment by smartphone shopping method, electronic commerce payment transaction payment method, in order to keep high technological or payment channel to attract customers.

The another element is how to deliver customer focused to feel differentiation, to fight for survival in the global market, a company needs to implement innovation function to cope effectively with the changes in the customers' needs. So, it is innovation element to the experiencing decline stage organization to help it to attempt to solve challenges in order to continue survive in possible . Innovation may include service, e.g. sale service, client service, delivery service etc. as well as product innovation ,e g. design change to the cup, mobile technological improvement, car style, design, engine, chargeable battery charge etc. or individual innovation, e.g. the fundamental assumption of operate culture changes, mindset of the top manger, CEO midset change and questioned by the capability of execution of the operations function to himself organization. How to innovate organizational learning, whether and what experience or knowledge applied in the operations function can be executed for the next innovation in order to renew the experiencing decline stage organization strategy to continue survive. So, innoviation is also one element to influence organization's survival.

● What can learn from the organizational life cycle theory?

The next element is whether organization can continue keep on learning element. In fact, there are many different factors to influence whether our organizational survival. They may include organizational internal factors as well as outside factors. In general organizations can control themselves internal to be better, but they can not control outside environment, because they can not predict when environments, e.g. economic environment, customer shopping desires, when new competitors enter market. According to organizational life cycle theory, during the firm's growth from inception to high growth, to maturity firm characteristics differ and the internal resources and capabilities of the firm develop.

Organizations encounter an unpredictable business environment which is constantly pressured by the changing effects of globalization , competition and technological advancement within the context of the knowledge economy ( Thoumrungroje & Tansuhaj, 2007).

reference

Thoumrungroje, A. & Tansuhaj, P. (2007). Globalization effects and firm perferences. Jounral of international business research, 6 (2),43-58.

During the first stage, any organization will up in a new business environment with much adaptation and try to develop a niche through, learning and innovative practices. Given the success of that survival, the organization becomes aggressive in the second stage in how to managing internal resources, effectiveness and efficiency, learning workflows, and corporate structure to accommodate the increased complexity of operations, policies are needed to implement in second stage. The third stage, business efficiency is the core, and the organization keeps resolving workplace problems and defining clear objectives of what to achieve short and long term in the business. In this stage, revenue is the key pursue aim or objective. The final stage, as a nature stge, the organization tends to maintain the business stability and spend time focusing on the status of how organizational structure, management departments cooperation strategy implementation and organizational culture in order to help the organization itself and the CEO leader himself/herself how to manage her/his organization successfully . So, whether the organization can continue keep on learning , it may be one important element to influence the organization to develop in business.

● How to develop organizations in growth stage?

Theoretical development of the organization life cycle has description of distinct stages of organization growth, Little attention to the dynamics of organization growth, such as how to grow the organization? Why can the organization grow rapidly ? I believe any organizations expect that they can grow rapidly, but the organizations expect that they can grow rapidly, but the question concens: What factors influence the organization can not grow rapidly, e.g. lacking proficient staffs, lacking high technological manufacture, lacking creating mindset or innovation, lacking suitable strategy implementation, strong new competitors enter, customers taste change etc. different organizational internal and external environment etc. factors. So, any organizations expect to grow rapidly, they need to solve the challenges to threaten their grow. For example, one mobile manufacturer only concentrates on manufacturing new technological smart phones products . So, at this stge, it ought pursue a niche strategy, presenting a very narrow smart phone product time, often a single smart phone product to a single smart phone user market. The new different design and function of smart phone products venture generally undertakes major and frequent smart phone product innovation. Major investments are needed to make in smart phone products development, robotic plant and manufacture robotic equipment is needed. So, this smart phone product manufacture firm expects to grow its smart phone share market rapidly. It must often need to innovate itself manufacturing technology , e.g. manufacturing robotics as well as designers need have good creative mindset to attempt to design any kinds of smart phones and change kind functions and smart phone design in ordeer to satisfy smart phone users' needs. When , it often have different kinds of new smart phone products design. I believe that it can grow itself smart phone productive speed rapidly, when it can attract many smart phone users to choose its any kinds of smart phone products to buy to use in preference . Hence, this smart phone manufacturing firm needs have good design mindset to attract smart phone consumers' attention if it expects to reach the growth stage in short time. If it feels that it has not increase smart phone customer number significant in this time, it may believe that its busines can not reach growth stage in the moment. Hence, the life cycle theory offers expected obstacles for each stage, which can help the firms to solve the problems and help them to attempt to find any useful strategies accordinglyly.

Hence, organizational life cycle can help any organizations to revise whether what obstacles can influence or threaten the organization itself can not grow easily, due to consumers demand have become more complex, as well as arket trends are harder to predict and competition is severe than ever, in order to survive companies, shoulf need to learn how to reach growth stage rapidly in order to raise its competitive ability to reach mature stage rapidly in short time. Becuase of the organization can reach mature stage from birth to young stage in short time rapidly. Then , it will have enough ability to avoid to reach decline stage rapidly. If it expects to continue survive or it does not need to review its organizational strategy ocnsequently. Because different stage, it will have different kinds of challenges to the organization will encounter, e.g. in the birth stage , challenges may include lacking cash flow, less customer number, without building attractive or famous product brand or image, in the young stage challenges, may include strategy is not suitable to implement effectively, customer growth speed is slow, product development or promotion challenges when the organization reachs mature stage, its challenges may include clients number begins decrease or loss old clients, product sale number begins reduction from the top level, customers feel its products are not attractive and they begin to choose to buy other similar feature of products to replace its products in market. Hence, when the organization can know whether it reachs which stage in its organizational life cycle model. Then, it may attempt to find the most suitable or the most reasonable strategy to implement in order to keep its long survival to stay on the mature stage more easily.

CHAPTER TWO

# Technology how influences economic recession or boom

COVID -19 disease how influence businesses may experience either growing life cycle stage or decline life cycle stage.

Nowadays, we are facing global economic recession period, since COVID 19 human mouth disease effect can bring economic crisis. Can it influence businesses feel difficult to adapt how global economic recession change after their decline life cycle stage? However, the effects of COVID 19 spreading will have wider implication , not just on how economies function, but also on how consumers behave, across china, Asia-pacific and around the world. Another effect of China;s economic rise is its influence in the adoption and adaption to new technological invention to manufacture , e.g. manufacturing robotic products had sold to China factories to replace workers to manufacturer products. It also will influence many China manufacturing workers lose jobs, when many China factories apply manufacture robotics to replace them in nowadays economic recession period.

Considering the adoption of online-offline shopping and home online office tasks, they are influenced by COVID-19 human disease influence, it also influences on regional travel in China, even global travel income is also reducing, because many travelers feel afraid to catch air planes to avoid to get COVID 19 human disease when they are sitting in close window airplanes by air . HOwever, COVID 19 also influences global consumer behavior changes to online shopping, because many people are afraid to enter crowd shops to avoid get COVID 19 human disease easily. So global shops will lose many visiting shop consumers, if they do not decide to attempt to open online stores to let customers to apply internet to buy their products. So, COVID 19 human mouth disease induced changes in consumer behavior. Shop online will be one new trend to influence young and old consumers make shopping from online stores. They will enquire whether the kind of product is worth to choose to buy by social media, e.g. facebook, online post . Hence, COVID19 human mouth disease may influence global economic recession, but it also brings e-commerce boom chance, when many consumers are fear to enter any crowd shops , when they need to stay long time in any shops. Then, they get COVID 19 human mouth disease chance will increase. Hence, it will influence many customers reduce to visit shops times, but it also creates online-shopping new business model . For example, China families are renewing their joy in home cooking. Onlins cooking videos are helping with the discovery od new recipes, new ways to create dishes , and new influences. So, opportunities are opening for more cleaning products, new ways to clean and new home hacks from online videos will bring global home consumers spend more time on their wellness or beauty routines ? So, COVID-19 disease also influences many families choose to cook dinner at homes at nght. Restaurants will lose many eating clients, because they are fear to enter restaurants to eat together to avoid to get COVID19 human mouth disease. But, it also creates home cooking products sale chance, e.g. rice cookers, dishes or any cooking tools because many families choose to cool at home. Hence, in some situation, economic recession will create new business chance , such as online store or rice cooker sale increases, they may be influenced in this COVID 19 human mouth disease occurrence environment.

Economic recession also influences business strategy changes. Many companies seem to be applying many aspects of a retrenchment approach , e.g. reduced fixed costs, narrower product offering, reduced staffs, but also there are some aspects of an investment approach which can be observed , because customers number will be influenced to reduce in economic recession environment. Companies have felt the robustness and quality of the approaches being applied had been allowed to decline. As a consequence of the challenges of a recession, urgent improvement have needed to be made because factories will reduce workers number to avoid salary expenditure spending more , but customers umber reduced in recession environment .

Hence, they will choose to buy manufacturing robotics to replace workers. If robotics can be improved to be proficient manufacture. Then, they won't need to buy many robotics to help them to replace to replace many workers to manufacture any products efficiently. So, manufacturing and improvement to robotics number demand may increase to any factories , e.g. vehicle manufacture, electronic products, e.g. computer hime cooking electronic products , e.g. rice cookers, heaters etc. products may be manufactured by manufacturing robotics. It creates the manufacturing robotic sale improvement quality chance in recession environment. It may impact on medium, or long term, it depends on how long time of recession. So, economic recession may bring robotic manufacture industry boom , when electronic products manufacturers need many improved robotics to replace workers in factries in order to reduce spending too much salaries expenditure in recession.

It is one external environmental factor to influence sudden manufacture robotic industry boom absolutely ,because electronic manufacturer's manufacturing robotic needs increases in recession environment. So, robotic manufacturers' strategy need to change , such as how to improve any manufacturers' needs in recession, e.g. manufacturing robotic product categories, market segments, geographic areas, core technologies, reliability , price, customisation, robotic manufacturing efficiency how to be improved of business.Change strategy to any manufacturing robotics manufacturers. So, recession may influence some kinds of manufacturing robotics' needs raise in robotic manufacturing market.

● How recession influences the role of advertising changes?

Advertising plays a key role in a dynamic economy. It may provide valuable information about products and services in an efficient manner, communicates client value, builds brand awareness and creates demand. However, when one country is experiencing recession, how it influences the country's businessmen spending on advertisement behaviors? Due to clients number reduces, a company usualy cuts come from the advertising budget than companies begin to cut back on advertiseing during an economic recession, they become less visible to the public because they predict clients number ought reduce next three months, even half year or one year. It depends on how long economt recession occurs. So, economic recession many impact any companies' advertising budget expenditure to be reduce . How much on the reduction on advertising budget expenditure, it depends on the company predicts how many clients number will reduce.However, due to advertising number reduces, it can influence consumer behavior changes indirectly.

In economic boom environment, consumers can watch to different kinds advertisement from television. Advertisement may bring positive alternative evaluation phase of biying decision-making process is bring exposed to buy several communication messages. In such an economic boom environment, any organizations may be clearly heard by the consumers, after any advertisement programs are broadcasted on television. Therefore, advertisemtn can persuade clients to choose to buy the kind of product after the kind of product advertisement is broadcasted from television absolutely.

However, when recession occurs, any companies; advertisement time is shortened , even number is reduced . Hence, they can not receive any client's positive or negative feedback immediately in short time afer advertisements are broadcasted from television . So, recession may influence advertisement time is shortened and number is rediced . On consequence, companies can not have any repsonse to know whether how market or customers' demand is changing to themselves products in shor time.

However, recession may bring worse advertisement effect to influence any businesses . On one hand, there is a negative economic recession environment because of the negative media reporting, these would be a decline in demand for the products and services and eventually companies would want to save more than they spend , But in the other hand, when the companies cut back advertiseing expenditures, they become less visible to public. Hence recession may influence many companies brand image will be lost, due to spending on advertisement expenditure wil reduce. Then, clients number may be influenced to reduce, because they can not watch the kind of product advertisment from television home often.

When one country is encountering recession, how are the various components of household consumption affected ? How is the impact of the recesion distributed across socio-demographic group? How does the recession compare to previous recessions? When book will boom? In fact, any country's recession may impact consumer behavior changes, it depends on these factors: age, race, education and wealth groups resulted in a decline in consumption inequality. The rich group is the " wealth effect influence group" when recession comes, it may influence their wealth reduces, so their enjoyment dsires will be influenced to reduce, e.g. purchase expensive cars driving enjoyment desires, purchase expensive house living enjoyment desires. If one rich person loses jobs , it may influence him to spend less time to drive themselves cars, so consumption of gasline will be influenced to reduce.

Economic theory ( e.g. consumer behavioral economic theory) predicts that when economic recession occurs, it will cause many businesses may experience decline cycle life stage rapidly, that link between income shocks and consumption has close relationship, such as rich person consumer group, if his income reduces, then he will buy less gas to drive himself car, even if he loses his job in recession environment, he will choose to sell his car to exchange cash. Hence, consumption may fall as a direct consequence of a fall in income induced by job loss, reduced hours or productivity and negative returns from assets, if there are long term changes to a household's econmic resource in recession environment. Hence, in recession environment, job loss or income reduction factors that may affect consumers and their shopping attitudes in the recession period. Otherwise, for low income group, recession may influence food consumption to low income consumer behavior changes to worse. Because low income person may reduce income ot lose job, then cheap food consumption will be influenced to worse to low income consumer group.

In recession period, if the food price is raised , due to the cost increase of food, it will lead to change in the reductin on quantity and type of food being purchase to low income food buyers. This may lead to a reduction in the quantity of food consumed and/or the substitution of high-priced food for cheaper food, which is often less nutritous and of worse quality. Hence in recession perios, low income food consumers will consider whether the kind of food price has how much increase or decrease. They won't consider the quantity of food consumed for maintaining energy balance and the quality of food consumed for maintaining ample intakes of protains, fats and micronutrients, such as vitamins, minerals and trace elements on food issue. So, if the kind of food price reduced in recession period, it ought may attract many low income food consumers number, even its food nutritious is worse. Hence, if the kind of meat price can be reduced in recession , the cheap types of meat consumption to low income consumer may be increased, even its nutritious is worse to compare the recession occurs before period.

On conclusion, in either economic recession or boom period, in general, consumer behavior will be influenced to change. Some products may be influenced to have higher sale in recession period, e.g. home electronic rice cookers , due to COVID 19 human mouth disease influenced many households choose to cook dinner at home at ight. Otherwise, some products may be influenced to have lowr sale., e.g. expensive cars sale in recession period, many high income people may lose jobs or reduce salaries , then it will influence their car purchase desires to be reduced. But if COVID 19 human mouth disease has medicine to kill this kind of disease. Then, economy will boom, many households will choose to go to restaurants to eat dinner. The, the electronic rice cookers sale number may reduce, when they reduce time to cook at home at night. Hence, it explains why economic recession or boom period may have impact to influence consumer behavior in behavioral economic view.

Applying business development strategy to raise the educational robotic manufacturer sale number in recession period

● What does business development strategy ?

An effecting business development strategy ought have these five steps: The first step is market analysis. Who are your clients , knowledge of your market? Second step is how to adopt for each penetration, your business needs to learn how to adopt for each group of clients, your first need to review your own capacbility. It is important that you are realistic and honest with yourselves over where clients truly sit, learn how to classify your clients into similar groups relative is the scale of the opportunity. Third step learns how to review your performance , market matrix to plot your results to help you determine your market penerstion. In addition, it will help you then discuss and consider various strategies for growth. By potting your clients you will get a sense of where your strengths and weaknesses are against the opportunity that total market offer.Fourth step learns how to consider alternative growth strategies on the market matrix. The final step , you need to consider these questions in order to decide whic is the most effective strategy for your business. For example, which model is the most ( least effective? Why? which model work best for line managers, HR are finance, why? How might we most effectively progress from one model to the most reasonable questions? ) Then, you will need to decide how to launch new services, new products, opening new markets, how accessing new geographic territories.

● How to apply business development strategy to help educational robotic manufacturers to enter traditional education market ?

Many thinkers concern robots that are used in manufacturing workplaces, homes, roads, hospitals and care centre aspect, but they don't feel robotics may be possible to apply on social service aspect, e.g. educational service aspect . In fact, robotic may have both functions. Industrial robotics, e.g.

manufacturing function as well as service robotics, e.g. professional robotcs, medical robotics, entertainment robotics, e.g. toy and education robotics and service robotics , e.g. personal and domestic robotics.

Educational robotic is on the birth stage in its industry life cycle. So, any educational robotic products will need time to persuade schools or any educational institutions to buy their products to assist teachers to teach students in classrooms. The question is how to apply business development strategy to help the educational robotic manufacturer to develop its educational robotic products to persuade educational clients to choose to buy ? I shall attempt to explain as below:

Due to educational robotic product is one new educational tool to assist any schools to buy to assist teachers to improve teaching service performance to let students to feel more learning satisfaction, so any educational robotic products must need time to introduce whether what it can bring schools benefits to let students and teachers to feel. When robotic can be popular to use on manufacturing, educational service industries aspect, e.g. warehouse , factory, shopping center, even restaurant's kitchen cooking robotics, office environment's accounting, law draft etc. clerical robotics may be invented to replace human 's general simple tasks. However, if future robots can be applied to educational aspect, e.g. classroom, school teaching students. Can educational robotic may assist or replace teachers to teach students in clasrooms? Will future teachers be replaced by teaching robotics . I shall attempt to explain whether it is possible that educational robotic can be developed to global educational organizations successfully as below:

Robotic technology has been invented to own " mind " ability, e.g. writing words, writing song, simple calculation tasks reading tasks . So, future robotics can be invented to own " mind " ability, when robotics' mind ability can be improved to own how to " communication" ability and " analytical" ability. Then can it be possible to apply robotics to do teaching tasks in classrooms, e.g. learning any books , the it applies the book's contents to analyze any "knowledge" in order to follow the logic mind to teach students in classroom. It is one major factor to influence any schools to explain why they need to buy any educational robotic in schools in any educational robotic product business development strategy. So, they need to find whether what their educational robotic strengths , any competitors won't own or their product weaknesses, they need to improve their educational robotic products in order to attract educational organizations to choose to buy.

Can future teaching robotics learn to do teacher individual same education tasks? It will be absolute competitive point to any educational robotic product manufactures. If it is true, can teaching robotics may be trained to exceed teacher individual teaching skill? It is another competitive point to any educational robotic product manfacturers. Is it ethic to apply robotic to teach students to replace teachers if teaching robotic can perform better teaching service to compare teachers? If the eduational robotic manufactuer can persude the school can accept eduational robotic ethic issue to assist or replace teachers to do teaching tasks, then it's sale chance will raise. So, ethic to educational robotic will be another factor to develop the educational robotic business. Can teachers be teaching robotic's teaching assistant role if teaching robotics can have teaching ability to teacher in the school? So, if the educational robotic manufacturer can persuade the school to feel that its products can be teacher's assistant to improve their performance to let students learn more easily. The educational robotics manufacturer may develop its product to sell in this educational market more easily, in business development strategy view.

All of these will be future any one educational robotic development challenges if they hope their products can sell more easily. They also need to know to let schools to know these disadvatages to their products to become advatanges in order to attract they to choose to buy their producte more easily. Such as what potential harmful consequences may come from the inventing of teaching robotics? What happends to important education moral, such as teacher or school privacy then robotic are starting to become an teaching tool to the school? Do such robotics hace any roght and responsibilities if the class has many students learning ability are influenced to poor or examination results are poor when the educational robotic has been bought to assist the school teachers to teach their students? Why does the school need to buy educational robotic to do teaching tasks? Any school organizations must need any one educational robotic product seller to answer any one of above questions, before they decide to buy their products. So, they must need to ensure teaching robotics will be used to help the school to teach students to learn more understanding to compare teachers only.

● Future educational robotic are applied on development teaching maths market

In the future, business development to educational robotic market may be teaching maths. I shall explain as below:

It is possible that students can use mobile robotic to learn mathematics subject to compare teachers more easily. Why? For example, young age from 4 to 14 age, they may apply mobile robotics to learn add, multiple, divided, simple math equation more understanding than math teaches. Robotic kints and apps is currently available on the maket for teacher of 4 to 14 age students,due to mobile , kits app price is cheap. So, they can be popular to be accepted by any primary schools , even in secondary schools, robotics may be applied to teach computer science, statistical methods subjects of one robotic kit for teach team of 2 to 3 students, short theory lessons , and tutorials to link theory and practice, realistic but affordable tasks linked with curricular subjects, teachers at ease with the robotic etc. So, future primary and secondary , even university teachers may need to choose the more suitable robot kit for their students,and carefully design where and how to use it and with which role.In fact, children will be possible to raise interest to learn when they can contact for any kind of teaching robotic to learn maths in classrooms together. So, teaching robotics may help 4 to 14 children students to raise learning interest instead of learning about ability.

In the future, robotic role is school may be one tool to engage the students as teachers role may be transfer base knowledge when teachers teach maths, geography, statistics, computer science subjects to promary , secondary even university students. This is one good example , whether what subjects robotic may be applied when it is invented to own human mind and anlaytical skill and communication ability. Robotics can perform more better to be applied to teach these subjects. It can let students to understand easily, e.g. understanding how to create equations that describe numbers a relationship understanding solving equations as a process of reasoning and explain the equations and inequalities in one variable, helping students to find different solutions, then best solves the problem , given the criteria and the constraints, helping students have more understanding how science knowledge is based upon logical and conceptual connections between evidence and explanations, even robotc can ask questions that can be investigated within the scope of the classroom, outdoor environment, and museums and other public faciltities with available resources and when appropriate frame a hypothesis based on observation and scientific principles. Even, robotics may help students to learn how construct, use and present oral and written arguments supported by evidence and scientific reasoning to support or refute an explanation or a model for a phenomenon, or robot can hep students to learn how obtain, evaluate and communicate information in 6-8 builds on k-5 and progresses to evaluating the merit and validity of ideas and methods, integiate qualitative scientific and technical information in written text with that contained in media and visual displays to clarify cliams and findings, helping students to anlyze data from texts to determine similarities and differences among several design solutions to identify the best characteristics of each that can be combined into a new solution to better meet the criteria for success, even helping students to learn how analyze data in 9-12 builds on k-8 and progresses to introducing more detailed statistical anslysis, the comparision of data sets for consistency and the use of models to generate and analyze data, analyze data using tools, technologies , and/or models e.g. computational, mathematical in order to make valid and variable scientific claims or determine and optinal design solution more easily than human teacher. So, there are human-made educaton machine advantage to students more than human teacher.

Educational robotic has been introduced as a powerful, in fact, flexible teaching / learning tool stimulating learns to control the behavior of tangible model using specific programming languages ( graphical, or textual and involving them actively in authentic problem -solving activities. Howeverm in future educational robotic development, it may be divided two separate categories as below:

Robotics as learning object: This first category includes educational activities where robotics is being studied as a subject on its own. It includes educational activities aimed at configuring a learning environment that will actively involve learners in the solution of authentic problems, facing on robotics -related subjects, such as robot construction, robot programming and artificial intelligence as well as robotic as learning tool: In the frame of this second category, robotics is proposed as a tool for teaching and learning other school subjects at different school levels. Robotics as learning tool is usually, seen as an interdisciplinary, project -based learning activity drawing mostly on science, maths, informatics and technology and offering major new benefits to education in genera at all levels. However, I believe the role of teacher is crucial for the successful industry of technological and innovations in classrooms, when robotics are been particiapted to any education tasks in classrooms. Schools can focuse on the training of prospective and in-service teachers in the use of robotics technologies through courses.

In future electronic learning environment, robotics can be participated, such as recognised their active participation in all sessions of the course and their creative involvement even in the theoretical parts introducing principles and methodology for designing robotic-enhanced projects, very much liked the activity-orientation of the educational content, acknowledged the central role of the e-workspace during the face-to-face meetings and beyond ehem in enhancing sense of community, acknowledged the potential of educational robotics as a teaching tool but also as a subject, in different disiplines , such as technology, informatics and engineerinfg, highly appreciated the opportunity to create their projects.
How to develop robotic in technological subjects on teaching, learning and educational aspect? Learners can be encouaged hen robotics participate actively in the learning process. Through robotic learners build something on their own, preferably a tangible object, that they can both touch and find meaningful. In robotic learners are invited to work experiments or problem-solving with selective use of available resources, according to their own interest, search and learning strategies. Robotics can help them to seek solutions to real world problems, based on a technological framework meant to engage students' movitation. So, when students can have control of specific robotc in a rich learning environment, the construction of robots and programs to control them the emphasis might move on interesting learning actiities in the frame of specific learning areas , such as science and technology. Thus, the design of robotic construction activities is associated with the fulfillment of a project aimed at solving a problem. In such a learning environment, learning is driven by the problem to be solved. To engage students in activities requiring to design and manufacture real objects, i.e. robotic structures that make sense for themselves and should devise activities that will encourage students to support in order experiment. So, robotic participation any science experiments, they may encourage students to create problem solving and combining interdisiplinary concepts from different knowledge areas,: science, mathematics , technology and research educational tasks, the role of students will change, when preparing a work with a programmable robotics studies experiment with simple programmable sobotics devices , e.g. a car-robot, motors, sensor etc. Students are asked to synthesize their finds and reach conclusions and solutions to the problem uner investigation. SO, robotic is educaional participation to any scientfic technological or research experiments, they may help students to work with creativity , imagination

and independence and finally organize the evaluation of the activity in collaboration ith studens. Also robotic participation to any technological or scientific research experiemtn, it also change teacher role . The teacher is such a constructist theoretical framework, like that teacher 's role that does not transfer ready knowledge to students, but rather acts as a organizer, coordinator and facilitator of learning for students. when educational robotics participate to any science or technological any research experiments, students may be organize the learning environment, raise the question , problem to be solved by students allow students to work with creativity, imagination and independence and finally organize the evaluation of the activity in collaboration with students. So, any educational robotic manufacturers must let their school clients to feel all these benefits which can bring to let students to raise learning abilty and learning interest to compare that are only taught by teachers, if they hope their educational robotics can be sold successfully in business development strategy view.

CHAPTER THREE

# Learning behavioral economy to solve social challenges

● Why do some social challenges may influence customers number ?

In our societies , we shall have different challenges to our every day. However, in general , the challanges seem that they do not have relate to influence businessmen profit, but in fact, these social challenges have relationship to influence business profit and clients number. I shall indicate some social challenges to explain why these social challenges may influence any business profit indirectly as below:

In investment or raving individual preference decision aspect, for some people , it may be interesting or fun to think cbout the best investments or the right health care plan. But, for other people, these choices are unpleasant, they may be persuaded to buy anythings, e.g. car, computer. So, if car seller can have persuasive methods to influence many people feel the health care plan or investment plan is not prefereable choices, driving car enjoyable feeling or material enjoyment is the most preference choice. Then I believe that the car seller's car selling number may increase, because some people greatly enjoy thinking about their pension and the best investment or health care insurance preferable decision, their decision had been influenced to choose to buy the car seller's cars. When they feel driving car enjoyable feeling is more important than future benefit.

Hence, in behavioral economy view, they had felt the driving car benefit is much to compare pension investment or health care insurance future benefit. The question is how to car seller can persuade these investment ot pension plan or health care preference decision individual to change purchase car driving decision>

I suggest that the car seller may have discount or cash coupon or installment payment method to attract them to consider , instead of advertisement promotion method. because this preference investment or pension saving or health care plan decision individual customer group will be more difficult to persuade them to choose to buy car immediately at this moment. Hence, if the car seller can not implement cheap car discount strategy, it will be difficult to attract this prefeence long term future benefit consumer to make purchase ca r decision easily. Because they think pension or investment or health care plan ce help them to bring long term future benefit, also it means that purchase car may only bring short term present benefit. It is general social behavioral consumption model to influence their purchase choice. Hence, I assume that general social long term future benefit product or service, e.g. insurance, investment , pension may influence th scocial shor tterm present benefit product , e.g. car consumer. It is the main reason, it can explain why car sellers can not persuade this long term future benefit consumers to make decision to buy their cars easily, when they have no enough money to spend to buy car and make investment, saving , medical care insurance , pension plan in the same time. They must need to make either purchase car or insurance etc. decision in our nowadays societies.

So, in behavioral economic view, it explains why consumer individual purchase choice behavior has relationship to himself/herself spending budget. I assume that it has two kinds of behavioral economic consumers. One kind if long term future economic benefit in preference more than short term present economic benefit, such as purchase car and investment or health care plan insurace saving term present benefit consumer, he / she considers to earn driving enjoyment at this moment is not preference than purchase insurance or investment future benefit decision . So, our society, any business will encounter these two kinds of behavioral consumer. They persuade either long term futuer benefit consumers or short term present benefit consumer to change himself/herself products or services more easily. Otherwise, such as if car seller can not implement coupon or cash reward or discount or installment cash payment strategy to attracr the long term future benefit consumer. Then, it will lose this group car customers number absolutely. So, it explains why businessmen need to learn consumer behavioral consumption model in order to increase client number more easily.

" Social welfare" usually measured by people's prefences, and it also focuses for the conventional economists, on how to maximize social welfare. What then is the task of behavior law and economics? Such as, this cate seller case example, whether what social welfare the car seller can bring to society when the individual decides to buy its car to drive or when the individual chooses to buy health care insurance or pension plan or investment . When he/she chooses to buy health care insurance or make pension plan or buys any companies‘ shares. Then, these investment service companies will bring what benefits to our society? So, instead of consumer benefit, we also need to consider whether the kind of product or service will bring what long term social benefit . However, I think that when global many people own cars, then many cars are driven on the roads, it will bring serious air pollution to influence our health. Then, when many people are got lung diseases by air pollution. Then, many people will need to pay more medical expense. It will be long term negative medical cost increasing expense to future us, but it also bring possible income for insurance firms, when many people plan to buy medical care plans when they feel air pollution will influence them to need to pay future medical expense. So, it seems that the effect on many people own cars and their driving behaviors will bring serious air pollution, but it will also create the health care medical insurance need to be increased due to many people feel air polluton will bring lung disease and they need to pay long time medical expensein the future long time in possible. So, many people driving behavior may bring air pollution, but it also bring medical insurance need increases in our society in possible. It means that air pollution may create medical insurance market develops in possible, such as most smokers say they would prefer not to smoke, and many pay money to join a program or obtain a drug that will help them quit.If many smokers forgive to smoke, then the medical care insurance need for smokers number may be influenced to reduce.

In social benefit view, medical insurance for smokers insurance will be influenced to reduce, due to many smokers forgive to smoke. Although, many smokers may get health, when they do not smoke, they do not pay to buy any cigeratte often, they can save more money, but cigeratte sellers and medical insurance service providers , their income must be influenced to reduce. Hence, when our society government's advertisement concerns smokers often smoke cigeratte, it may bring poor drug health or many cars air pollution, these two messages may influence or dissuade many smokers forgive to smoke or many people do not buy cars. They choose to catch public transportation, or owning car people who do not often drive cares, then car gas or fuel suppliers income will be influenced to reduce, due to many car owning people do not often drive cars or many people do not choose to buy cars. Then, car sellers' income wil be influenced to reduced.

Moreover, in long term social influence, when many people do not feel lung disease . Then, the medical care insurance need will also influenced to reduce.

It may bring insurance industry develops in difficulty for lung dissease medical care insurance. So, it explains why consumer behavior may also influence our social economic development in long term . They have cause and effect close relationship. When many consumers individual forgive or dislike to do the behavior in habit, e.g. driving car behavior or smoking behavior. Then, it will influence car seller market and cigeratte seller market to be poor in any countries , even global market.

Hence, in our society, when one individual feels that he.she has individual challenge, it may be economic or emotion or health problem, such as smoking influences health case, driving influences air pollution case. These both kinds of individual behavior may influence the individual may need to spend money for lung disease if he/she has continue smoking habit every day or he/she often drives car . Then, the individual will seek methods to solve these possible occurrence of problems before they do not occur. As it occurs in the natural environment, e.g. air pollution or lung disease is caused by cars or smoking. When individual begins feel these negative effect may case, if he/she continues to do smoking or dirving car behavior. He/she will begins to find methods to solve problem, problem solving is defined as the self-directed cognitive -behavioral process by which an individual , couple or group, such as smokers and drivers group in our society, they attempt to identify or disciver effective solutions for specific problem encountered in everyday living. More specifically, this cognitive -behavioral process ( a) makes available a variety of potentially effective solutions for a particular problem and (b) increases the probability of selecting the most effective solution from among the various alternatives ( D'Zurilla & Gold field 1971).

reference

D' Zurilla, T. J. & Goldfield, M.R, (1991). Problem solving and behavior modification, Journal of abnormal psychology, 78, 107-126.

As this definition implies social problem solving is conceived as a conscious, rational, effortful, and purposeful activity. Depending on the problem solcing goals, this process may be aimed at changing the problematic situation for the better, reducing the emotional distress that it produces or both.

Hence, it implies that when any one feels he/she will have individual problem, e.g. health problem , economuc problem,emotion problem. He/ she will avoid to continue to do the kind of behavior often every day ,e.g. smoking behavior or driving car behavior .When our society has many people make to forgive to do above themselves behaviors, such as smoking or driving habit. Then, it will influence cigeratte sale number and car sale numner to be reduced. So, when our society has any consumer groups, they forgive to do themselves behaviors in habit. Consequently, the kind of product seller or service provider may lose man customers. So, in our society , when one kind of product or service consumers , their habital behaviors are changed to reduce, then it may influence the kind of product sellers or service providers their income or clients number to be either decrease or increase. On conclusion, it explains that why social behavior has close relationship to influence business income or clients number in our societies.

CHAPTER FOUR

# Can space tourism influence economic development

Earth tourism recession how brings space tourism development
However, green or nature tourism strategy may include these elements : Quality, tourism should have an impact on the quality of life for all members of the tourist process, exploitation of nature resources should be optimal and ensure their generation, balance, distribution of benefits among participants in the tourist process must be fair. So, future any kinds of green or nature tourism will need have these features in order to attract many travelers to visit any countries' green lands, e.g. they may rent cars to travel to green lands. So, developing attractive green lands will be one kind new travelling trend for green tourism in global future travel market.
There are two types of models that contribute to the better understanding of future tourism industry development, explanatory model refer to factors that cause development growth. For example, whether the travelers feel necessary to travel to different destinations, very often nice landscapes and sightseeing, pescriptive modes ( e.g. life clcle explanations, physical models) examines tourism from what appears on ground e.g. large hotels facilities etc. Hence, any kinds of tourism leisure must need build these both models in order to attract travelers to choose to buy the tourism package from the travel agent more easily. It is important tourism leisure element to any one travel agent's tourism service package if it hopes to develop its tourism service success. So, the expansion of the tourist region over the natural boundaries of the city centre that occured in the first place as a result of the growth of tourism demand, is the end causing this very expansion to continue.

Butler (1980) involves a six stage evoluation of tourism, namely explanation, involvement, development, consolidation, stagnation, and post-stagnation. The last stage is further characterized by a period of decline, rejuvenation or stabilization. The applicability of the model to a given area has been assessed and judged of a tourist destination's development matched the six phases conceptually described by Butler

reference

Butler, R.W. (1980). the concept of a tourist area cycle of evolution: Implications for management of resources. Canadian Geographer, 24, 5-12.

Hence, our tourism industry is facing decline life cycle stage because COVD 19 human mouth disease has influenced many travelers feel fear to catch airplanes to travel, even they also feel to contact the potential COVD 19 human mouth disease people when they arrive the country , they feel that they may contact these sick people, instead of airplanes. So, this kind disease had influenced many travel agents reduce tourism service package number , due to many travelers' tourism leisure activities will reduce, due to travelers number reduces, they only carry cargos to transport to replace travelers COVD 19 disease influence our tourism industry is experiencing decline life cycle stage nowadays. Unless, COVD 19 human mouth attacking to lung disease can be treated by new medicine invention . Otherwise, tourism industry can not re-grow to mature life cycle stage easily.

The most used framework for examing stagnation and possible decline in tourism destinations has been tourist area life cycle model ( Butler, 1980). The model has been operationalized frequently in the tourism lierature. It includes series of stages in tourism development, leadning eventually to the stagnation and post-stagnation stages. When a nature destination can either decline, however, it does not offer a systematic explanation of hoe tourism destination might avoid decline . Such as COVD 19 human mouth disease may influence travelers feel fear to catch air planes. So, even the country has beautiful nature scene to attract people to travel, althoug it is a nature attractive destination, but due to COVD19 disease occurs, it may influence this country's this nature attractive destination to enter decline life cycle stage at this moment.

Hence, tourism industry's life cycle stage , sometime it can be influenced by non predicted factor, such as COVD19 disease factor, it can influence travelers' travelling desire to be reduced suddenly from 2019 , due to they feel afraid to catch air planes to avoid to get this kind COVD 19 human mouth disease to bring lung disease when they are sitting in closed window inside air plane environment. So, COVD 19 human counth disease causes global tourism industry is facing serious decline life cycle stage. The question is that any one does not know when this kind COVD 19 disease will be treated by new medicine invention, so if this kind COVD 19 disease still can not be killed by new medicine invention, then it will continue to influence global tourism development to be improved , even any nature attractive scenes, they can not persuade any travelers to catch air planes to visit any countries to travel easily. But, however, we still need to keep our natural environment to prepare future COVD 19 diease disappears , e.g. parks are important places for the protection of ecological systems and natural resources as well as for the provision ot recreational and tourism opportunities for the public. Then, nature or green tourism can be continue to develop to attract many travelers to travel after COVD 19 disease disappears in the future.

● What are the characteristics of birth life cycle stage to tourism industry ?
Butler , R.W. (1980)'s model begins with a discovery and exploration or birth stage in which a location is discovered by a small, select group of people as a place with desirable assets often, this discovery is nature population who may see the perceived assets. As just ordinary aspects of their environment or local culture. The early tourists have very little support in the form of amenities, and typically, this is preferred and is part of a location's of being undiscovered. The early tourists, therefore rely heavily on and interact frequently with the residents of the region. This small group of early tourists is largely in dependent and shares information about a destination by word of mouth or by select affinity groups. Over time, as more people are introduced to the destination, the number of visitors begins to increase. So " word of mouth" will be traveler information to persuade them to make travelling destination choices in the tourism industry beginning. It is tourism industry's birth life cycle stage characteristics . However, internet invention can let any one see any countries' scene photos, so it is one kind of good advertisement method to introduce any countries' scene, instead of travelling magazine in tourism growth and maturity life cucle both stages.

Moreover, space tourism is at the birth life cycle stage. It needs travelers feel interest to travel space, if this kind space tourism service providers hope to implement their any space journeys in success. These factors may influence its development succeeds. Nowadays, its target market is wealthy travelers group, wealthy individual are needed, as they serve as the main consumers for space tourism . For space tourism to succeed there must be enough demand from those who are able to afford to expensive ticket. To date there have only been seven commercial space travelers, or space tourists, although they prefer to be called space flight participant, as they see themselves as pioneers and adventers as opposed to ordinary tourists. So, any future space tourism that price must need to reduce to general public, e.g. ordinary income level people, they can spend, if space tourism hopes to reach from stage stage rapidly. So, space tourism is still far to mature stage.It depends on whether how long time its any space journey ticket price can be reduced to any one can pay. So, when its customer target is not only wealthy travelers, many ordinary or common income level people, they can pay to any one space jounrney. It may mean to reach growth life cycle stage.

● What characteristics to space tourism growth stage?

When human space tourism of commericalization of activities in outer space can bring these feeling to let any one space traveler feels then, it may mean that it can reach growth stage, such as they may feel their any space journeys may bring positive impacts that outer. Space recreation can produce, in order to come up with space tourism, exploring and untravelling the hidden anystories of the space are needed. Also they can feel need drastically broadens and enrichs human's technical awareness and constructive knowledge need from any one space tourism journey package. When space tourism reachs mature life cycle stage? What its characteristics are? When any one space travelers can feel that not only earth based attractions that simulate the space experience , they must need to catch airships to experience this different tourism experience, such as space theme parks, space training camps, virtual reality facilities , space hotels ( skotel), multimedia interactive games and tele robotic moon rovers controlled from earth, but also parabolic flights, lasting up to three days or week long stay at floating space hotel, including participatory educational ,as well as sports competitions ( i.e. space olympics). Hence, above these will be nay space tourism development. It can reach mature life cycle stage characteristics when any one can feel the real travelling mouth to compare to travel our earth anywhere, they can not find that they feel space tourism may be same to our earth's holiday ( need to rela) or cultural ( know different places or specialized tourism, e.g. expectations of adventures , even space scientists discover new experiences to expectations of adventure or get more information, scientific interest feeling. Then, at this moment, we can call space tourism has reached the mature stage. However, I believe that to develop space tourism in success. We must need to control space tourism ticket price to be reduced to general low income people. They may spend budget level. So, ticket price may be one major factor to influence future space tourism growth when it can reach mature stage. Also, it mean that whether space tourism may become another kind of popular tourism lesiure activities to use. It depends on ticket price factor, instead of its any space tourism trip arrangement factor. So, any one space tourism service provider must need long time to spend in order to implement its different strategies, e.g. ticket price, space trip arrangemet to achieve its their space tourism to achieve its their space tourism different destination package in success if they hope their future space tourism business can grow up in short time.

CHAPTER FIVE

# Robots How Help economic development

Robots How Improve Employees Performance

Assistance to reduce teaching work workload or psychological pressure to teachers in developing countries‘ schools

Another vital area benefiting from innovative technologies like AI is education. Advanced technologies can enhance how we learn, teach and perform tasks. In most developing countries, schools lack experienced teachers and resources to enhance students’ knowledge. As a result, many students still have to walk long distances to get to the nearest school, which has created education gaps, especially in rural areas. AI tools such as personalized learning assistants can simplify learning by making tutoring services and learning materials accessible to all students, wherever they are. Machines can be automated to help students learn basic concepts without a tutor, which companies like Carnegie Learning are working on. This would allow students to learn at any time from anywhere. With AI, education is made easy and accessible to more people.

The initial usage of AI in developing countries has been at a micro level -- solving small, specific problems in a defined industry. As machine learning advances and there is a higher utilization of AI, we will see more complex issues being targeted and resolved. When duly adopted, AI can positively impact future developing countries people everyday lives not just in disaster intervention, education, health care and agriculture but can also help in mitigating poverty, malnutrition and pollution. Especially, in developing nations, to leverage AI’s true potential and create a snowball effect. Startups are defining a holistic and humanitarian approach to building more sophisticated, AI-ready societies. Stakeholders in the AI landscape should understand the strengths and nuances of the developing world as well as the limitations of AI and create localized solutions and applications.

Why does smart phone help developing countries communication ?
Internet Seen as Positive Influence on Education but Negative on Morality in Emerging and Developing Nations. Internet access differs substantially across the 32 emerging and developing countries polled, with the lowest rates of internet use in South Asian and sub-Saharan African nations. Within countries, computer owners, young people, the well-educated, the wealthy and those with English language ability are much more likely to access the internet than their counterparts. To access the internet, people increasingly use smartphones rather than more cumbersome fixed landline connections and computers. Around the world, both smartphones and basic-feature phones alike are used for sending messages and taking pictures.

In fact, many developing countries young people, students are popular to use smart phones for internet usage aim, instead of communication. Moreover, many developing countries working people are also popular to use smart phones for any working usage in their working time , even non working time any time. So, smart phones (AI) phones will be important communication or leisure tools to developing countries people in the future. Unless, it is one day, scientists can develop another new communication tool to replace smart phones. So, artificial intelligence will be important to influence developing countries people , how to improve or bring positive learning attitudes to students in their daily learnnng lifes. as well as how to raise developing countries people, how to raise working people efficiency or improve performace in their daily working lifes. So, AI may bring positive learning or working attitudes to developing countries working people and students both.

The Positive Impact of Mass Media in Developing Countries
Radio, newspapers, television, Internet, social media, etc., all of these are forms of mass media. Each of these outlets has the capability of bringing information to thousands of people with one device. While in some communities it is easy to take advantage of these communication outlets such as television and Internet access, not everyone has access to such outlets. Radio is one of the most common forms of mass media in developing countries because it's affordable and uses less electricity than many other forms of mass media, but only approximately 75 percent of people in developing countries have access to a radio, and roughly 77

percent of people in rural areas have access to electricity.

For developing countries that have implemented forms of mass media in their communities, there have been numerous positive outcomes are influenced to impact developing countries mass media by artificial intelligence as below:

When AI is participated to developing countries mass media, it can influence any radio, television audiences raise more attention to each other through social media platforms such as Facebook and Twitter and create, organize and initiate street protests and campaigns. Furthermore, having access to social media in developing countries, people are able to connect to those that they usually wouldn't have the chance to talk to. Moreover, AI Provides educational opportunities- In many countries, the division between local and national languages as well as issues of literacy can make communication difficult. With the use of mass media, a bridge can be built between these two gaps. In India, there is a radio station that provides information in local languages and respects local culture and traditions. One of the main ways is to create public awareness of what is going on with businesses and government officials. The media plays an important role in giving people the opportunity to act against injustice, oppression and misdeeds that they otherwise wouldn't know about. Information on available healthcare, a mass radio broadcast was sent out encouraging parents to seek treatment at local healthcare facilities for their sick children. With this mass outreach on healthcare, the encouragement of people to take their children to healthcare facilities saved thousands of lives. This easy way of encouraging others and bringing awareness about certain diseases was made possible through a simple radio broadcast. Finally, when AI is particiapted to media, it may bring many social issues to life that otherwise would remain unknown to many people. In developing countries and communities like Burkina Faso, when the radio broadcast was released about malaria, diarrhea and pneumonia, people were educated and moved to action and knew to take their children to healthcare facilities for preventative care. As it is seen, having access to different media outlets is vital for those in developing countries. Here are three ways that those in developing countries can implement mass media to help their people and communities.

When AI is participated to any internet radio or internet newspaper mass online listening or reading channel. It can provide online radios or newspapers in public places- By providing online radios and newspapers in public areas it gives community members to access news, information and emergency warnings. Even though radios can be on the cheaper side, there are still many people that can't afford to have a radio in their home. By providing one in a local place, not only would it better educate the community members but also it will bring the community together. So, it can make media outlets a two-way platform- Creating a two-way platform between the community and those who are behind the radio stations, newspapers or broadcasts makes the community feel involved and that their voices are being heard. An organization called Soul City in sub-Saharan Africa is showing how well two-way platforms work by engaging their listeners and having them contribute thoughts and ideas about complex issues. Because developing countries radio listening audiences or newspaper readers are popular to accept computer online radio listening channel or online newspaper reading channel to replace traditional paper newspapers or radio machines. So, AI may raise their listening news or reading news leisure feeling from online mass media channel in the future.

● Why do developed countries need to develop AI

Artificial intelligence, or AI, is driving massive shifts across the globe, and every day more questions arise. What impact will AI have on the workforce and how can we prepare for it? How can we encourage economy-boosting and job-creating technologies? How can we ensure that AI will be implemented ethically and with minimal bias? How will society benefit? For developed country, such as US example. None of the US, Israel and Russia have a formal national AI policy yet. Private sector companies such as Google, Amazon and Apple and the US department of defence are driving the bulk of AI investment in the United States. Though Israel does not have a specific policy, it is keenly focused on AI and has seen the number of AI start-ups triple since 2014.

Developed country may learn whether what weakness it is lacking when it does not continue to develop AI from one another developed country. Which countries are approaching AI most effectively, and to what degree is there opportunity for greater international collaboration? It may be too early to tell; however, when analyzing the best practices of existing national AI policies, there is much that can be learned. These are the specific areas to consider. When one developed country continue to develop or research AI, it may bring these benefits as below:

On gathering Data aspect, from self-driving vehicles to smart cities, data is the driver behind AI. Innovation in the United States is limited without a national strategy that answers questions about protocol and ownership. France and Denmark, on the other hand, are opening government data. France is hosting troves of centrally collected public and private data that it plans to make available as part of its strategy. Conversely, by taking a restrictive position on issues of data collection (as indicated by the implementation of General Data Protection Regulation), the EU is putting manufacturers and software designers at a disadvantage while balancing the demand for privacy. On raising technologica talent aspect, the demand for AI talent far outweighs the available supply. As a result, almost every nation's strategy addresses talent development. Canada's AI strategy is distinct in that it primarily focuses on research and talent strategy. The country boasts AI degree programmes and is building a $127 million research facility in Toronto. Companies like Facebook and my own company, Uptake, are investing in Canada to access this talent pool. On AI legal technological innovation aspect, a whole host of legal questions swirl around AI. The country is developing a bill for AI liability that will be ready in March 2019. The government hopes the legal framework will attract investors by providing a simple, comprehensive guideline to enable the broad use of AI systems. So, when the developed country applied AI technology to assist any lawyers to work, then AI can help them to reduce the workload to draft any legal documents more easier. So, any developed countries lawyers' draft legal documents time must reduce if the developed countries lawyers accept to apply AI to assist their legal works. One of the great promises of AI is its potential for improving quality of life. But without the right planning and oversight, we risk exacerbating problems of inequality or marginalizing groups of people. As an example, India's AI strategy is focused on leveraging the technology not only for economic growth, but also for social inclusion.

AI may bring what benefits to developed countries

From SIRI to self-driving cars, artificial intelligence (AI) is progressing rapidly. While science fiction often portrays AI as robots with human-like characteristics, AI can encompass anything from Google's search algorithms to IBM's Watson to autonomous weapons. Artificial intelligence today is properly known as narrow AI (or weak AI), in that it is designed to perform a narrow task (e.g. only facial recognition or only internet searches or only driving a car). However, the long-term goal of many researchers is to create general AI (AGI or strong AI). While narrow AI may outperform humans at whatever its specific task is, like playing chess or solving equations, AGI would outperform humans at nearly every cognitive task.

Why research AI safety? Would AI bring war when AI is continued to develop by developed countries? In the near term, the goal of keeping AI's impact on society beneficial motivates research in many areas, from economics and law to technical topics such as verification, validity, security and control. Whereas it may be little more than a minor nuisance if your laptop crashes or gets hacked, it becomes all the more important that an AI system does what you want it to do if it controls your car, your airplane, yoür pacemaker, your automated trading system or your power grid. Another short-term challenge is preventing a devastating arms race in lethal autonomous weapons.

In the long term, an important question is what will happen if the quest for strong AI succeeds and an AI system becomes better than humans at all cognitive tasks. As pointed out by I.J. Good in 1965, designing smarter AI systems is itself a cognitive task. Such a system could potentially undergo recursive self-improvement, triggering an intelligence explosion leaving human intellect far behind. By inventing revolutionary new technologies, such a superintelligence might help us eradicate war, disease, and poverty, and so the creation of strong AI might be the biggest event in human history. Some experts have expressed concern, though, that it might also be the last, unless we learn to align the goals of the AI with ours before it becomes superintelligent.

There are some who question whether strong AI will ever be achieved, and others who insist that the creation of superintelligent AI is guaranteed to be beneficial. At FLI we recognize both of these possibilities, but also recognize the potential for an artificial intelligence system to intentionally or unintentionally cause great harm. We believe research today will help us better prepare for and prevent such potentially negative consequences in the future, thus enjoying the benefits of AI while avoiding pitfalls.

How can AI be dangerous when developed countries continue to develop AI to become weapon to replace soldiers?

Most researchers agree that a superintelligent AI is unlikely to exhibit human emotions like love or hate, and that there is no reason to expect AI to become intentionally benevolent or malevolent. Instead, when considering how AI might become a risk, experts think two scenarios most likely:

The AI is programmed to do something devastating: Autonomous weapons are artificial intelligence systems that are programmed to kill. In the hands of the wrong person, these weapons could easily cause mass casualties. Moreover, an AI arms race could inadvertently lead to an AI war that also results in mass casualties. To avoid being thwarted by the enemy, these weapons would be designed to be extremely difficult to simply "turn off," so humans could plausibly lose control of such a situation. This risk is one that's present even with narrow AI, but grows as levels of AI intelligence and autonomy increase.

The AI is programmed to do something beneficial, but it develops a destructive method for achieving its goal: This can happen whenever we fail to fully align the AI's goals with ours, which is strikingly difficult. If you ask an obedient intelligent car to take you to the airport as fast as possible, it might get you there chased by helicopters and covered in vomit, doing not what you wanted but literally what you asked for. If a superintelligent system is tasked with a ambitious geoengineering project, it might wreak havoc with our ecosystem as a side effect, and view human attempts to stop it as a threat to be met. So, a super-intelligent AI will be extremely good at accomplishing its goals, and if those goals aren't aligned with ours, we have a problem. You're probably not an evil ant-hater who steps on ants out of malice, but if you're in charge of a hydroelectric green energy project and there's an anthill in the region to be flooded, too bad for the ants. A key goal of AI safety research is to never place humanity in the position of those ants.

Why the recent interest in AI safety ?

Stephen Hawking, Elon Musk, Steve Wozniak, Bill Gates, and many other big names in science and technology have recently expressed concern in the media and via open letters about the risks posed by AI, joined by many leading AI researchers. The idea that the quest for strong AI would ultimately succeed was long thought of as science fiction, centuries or more away. However, thanks to recent breakthroughs, many AI milestones, which experts viewed as decades away merely five years ago, have now been reached, making many experts take seriously the possibility of superintelligence in our lifetime. While some experts still guess that human-level AI is centuries away, most AI researches at the 2015 Puerto Rico Conference guessed that it would happen before 2060. Since it may take decades to complete the required safety research, it is prudent to start it now.

Because AI has the potential to become more intelligent than any human, we have no surprise way of predicting how it will behave. We can't use past technological developments as much of a basis because we've never created anything that has the ability to, wittingly or unwittingly, outsmart us. The best example of what we could face may be our own evolution. People now control the planet, not because we're the strongest, fastest or biggest, but because we're the smartest. If we're no longer the smartest, are we assured to remain in control?

A captivating conversation is taking place about the future of artificial intelligence and what it will/should mean for humanity. There are fascinating controversies where the world's leading experts disagree, such as: AI's future impact on the job market; if/when human-level AI will be developed; whether this will lead to an intelligence explosion; and whether this is something we should welcome or fear. But there are also many examples of of boring pseudo-controversies caused by people misunderstanding and talking past each other. When one developed country continue to develop AI, can itself country's all factories workers will lose jobs, due to AI can replace them to do simple works in factories, or any public transport drivers, e.g. bus drivers, ferry , tram, train drivers, they will lose jobs, when AI ( non manual driving drivers ) can replace all public transport drivers. So, some occupations will lose if developed countries continue to develop or research AI to replace human to do some simple jobs, such as some cooking jobs can be done by AI. So, it is possible that future cookers won't be needed, because AI cooking skills may be better than them to cook any good taste chinese or western food in restaurants. If you drive down the road, you have a subjective experience of colors, sounds, etc. But does a self-driving car have a subjective experience? Does it feel like anything at all to be a self-driving car? Although this mystery of consciousness is interesting in its own right, it's irrelevant to AI risk. If you get struck by a driverless car, it makes no difference to you whether it subjectively feels conscious. In the same way, what will affect us humans is what superintelligent AI does, not how it subjectively feels.

In fact, AI may be make any brokers jobs in financial market. the main concern of the beneficial-AI movement isn't with robots but with intelligence itself: specifically, intelligence whose goals are misaligned with ours. To cause us trouble, such misaligned superhuman intelligence needs no robotic body, merely an internet connection – this may enable outsmarting financial markets, out-inventing human researchers, out-manipulating human leaders, and developing weapons we cannot even understand. Even if building robots were physically impossible, a superintelligent and super-wealthy AI could easily pay or manipulate many humans to unwittingly do its bidding. So, future brokers will be replaced by AI, when AI can be made to own financial brokers‘ analytical mind to make more accurate whether the share price will rise up or fall down to compare human financial brokers' analytical mind. The robot misconception is related to the myth that machines can't control humans. Intelligence enables control: humans control tigers not because we are stronger, but because we are smarter. This means that if we cede our position as smartest on our planet, it's possible that we might also cede control.

Not wasting time on the above-mentioned misconceptions lets us focus on true and interesting controversies where even the experts disagree. What sort of future do you want? Should we develop lethal autonomous weapons? What would you like to happen with job automation? What career advice would you give today's kids? Do you prefer new jobs replacing the old ones, or a jobless society where everyone enjoys a life of leisure and machine-produced wealth? Further down the road, would you like us to create superintelligent life and spread it through our cosmos? Will we control intelligent machines or will they control us? Will intelligent machines replace us, coexist with us, or merge with us? What will it mean to be human in the age of artificial intelligence?

Why do developed countries people need AI ?

Why do we assume that AI will require more and more physical space and more power when human intelligence continuously manages to miniaturize and reduce power consumption of its devices. How low the power needs and how small will the machines be by the time quantum computing becomes reality? Why do we assume that AI will exist as independent machines? If so, and the AI is able to improve its Intelligence by reprogramming itself, will machines driven by slower processors feel threatened, not by mere stupid humans, but by machines with faster processors? What would drive machines to reproduce themselves when there is no biological incentive, pressure or need to do so?

Who says superior AI will need or want to have a physical existence when an immaterial AI could evolve and preserve itself better from external dangers. What will happen if AI developed by competing ideologies, liberalism vs communism, reach maturity at the same time, will they fight for hegemony by trying to destroy each other physically and/or virtually. If AI is programmed to believe in God, and competing AI emerges programmed by muslims, christians or jews, how are the different AI's going to make sense of the different religious beliefs, are we going to have AI religious wars? What if the "powers that be" greatest fear is the emergence of a super AI that police's and rationalizes the distribution of wealth and food. A friendly super AI that is programmed to help humanity by, enforcing the declaration of Human Rights (the US is the only industrialized country that to this day has not signed this declaration) ending corruption and racism and protecting the environment.Most benefits of civilization stem from intelligence, so how can we enhance these benefits with artificial intelligence without being replaced on the job market and perhaps altogether?

Key to the process of machine learning are neural networks. These are brain-inspired networks of interconnected layers of algorithms, called neurons, that feed data into each other, and which can be trained to carry out specific tasks by modifying the importance attributed to input data as it passes between the layers. During training of these neural networks, the weights attached to different inputs will continue to be varied until the output from the neural network is very close to what is desired, at which point the network will have 'learned' how to carry out a particular task. A subset of machine learning is deep learning, where neural networks are expanded into sprawling networks with a huge number of layers that are trained using massive amounts of data. It is these deep neural networks that have fuelled the current leap forward in the ability of computers to carry out task like speech recognition and computer vision.

In conclusion, when developed countries continue to develop AI, it may bring positive advantages to bring raising productivies, or efficiencies, but it may also raise unemployment ratio to any low skill or low knowledge jobs in ther societies. However, human future society will need to change to be better to raise our living standard. But AI is one kind the best choice tool to achieve this aim in our future, so I agree developed countries continue to develop or research AI to be the super -human machine.

Reference

A. Castano et. al. " Automatic detection of dust devils and clouds at Mars" Machine vision and applications, Oct. 2008, vol. 19, no 5-6, pp. 467-482.

Accenture, " Why artificial intelligence is the future of growth"(2017) <http://www.accenture.com/us-en/insight-a rtificial-intelligence-future-growth>.

D. Schedidt , Unmanned Air Vehicle Command And Control, Handbook Of Unmanned Air Vehicles, Springer-Verlag, 2014. Facebook (AI) Research

Available at https://research.facebook.com/ai, research at google, machine intelligence available at http://research.google.com/pubs/machineintellige nce.html; micro soft research-machine learning and artificial intelligence available at http://research.microsoft.com/en-us/research- areas/machine-learning-ai.aspx.

International Federation Of Robotics, 2016. IFR press release world robotics report. IFR, org . 29 Sept. Accessed Feb. 01, 2017. http://www.ifr.org/news/ifr-press-release/world-robitics report -2016-8321.

K, Fedra , "GIS and environmental modelling" in environmental modelling with GIS, edited by M.F. Goodchild.B.O. Parks and L.T. Steyaert, Oxford University press, pp. 35-50, 1994.

Keynes, J.M. (1933). Economic possibilities for our grandchildren (1930). Essays in persuasion, pp.358-73.

Mckinsey & Company (2013, May). Disruptive technologies: Advices that will transform life, business and the global economy , USA.

Ministry of economy, trade and industry, Japan, 2015, Japan's robot strategy. Ministry of economy, trade and industry.

Ray Kurzweil , The age of spiritual machines (1999) is cited numerously through this chapter: Kurzweilai.net http://www.kurzweilai.net

Rich, Elaine & Knight, Kevin, Artificial Intelligence Second Edition, 1991, New York; Mc-Graw-Hill.

Artificial intelligence bank service working environment

Focus on outcomes not technology.Artificial Intelligence: Waiting to be unleashed? The Insider Column - When Digital Transformation misses. Are you meeting the demands of the new digital consumer? Will your legacy mindset compromise your digital competitiveness? Can artificial intelligence create online remote office new business service market in global ?

The Future of Artificial Intelligence In The Workplace:

Is AI going to displace workers or come as a benefit to them?

Is AI going to displace workers or come as a benefit to them? Getty

Smart technologies aren't just changing our homes; they're edging their way into their numerous industries and are disrupting the workplace. Artificial

Intelligence (AI) has the potential to improve productivity, efficiency and accuracy across an organization – but is this entirely beneficial? Many fear that the rise of AI will lead to machines and robots replacing human workers and view this progression in technology as threat rather than a tool to better ourselves.

With AI continuing to be a prominent online office service business to replace human actual office working environment, businesses need to realize that self-learning and black-box capabilities are not the panacea. Many organisations are already beginning to see the incredible capabilities of AI, using these advantages to enhance human intelligence and gain real value from their data. As there is increasing evidence demonstrating the benefits of intelligent systems, more decision-makers in the boardroom are gaining a better understanding of what AI can really offer. Research conducted by EY explains "organizations enabling AI at the enterprise level are increasing operational efficiency, making faster, more informed decisions and innovating new products and services." Can articial intelligent technology create remote office working environment to replace our traditional actual office work environment ? Can we do not need to go to office to work , when any office staffs ,e.g. managers, clerk, etc. they can apply artificial intelligent technology and online technology to work at home, such as remote office working environment ?

The first companies employing AI systems across the board will gain competitive advantage, reduce cost of operations and remove head counts. Whilst this may be a positive from a business perspective, it is obvious why this a worry for those working in roles at risk of displacement. The introduction of these technologies will likely trigger an issue with unions and job security due to the substantial operational changes. Although AI will affect every sector in some way, not every job is at equal risk. PwC predicts a relatively low displacement of jobs (around 3%) in the first wave of automation, but this could dramatically increase up to 30% by the mid-2030's. Occupations within the transport industry could potentially be at much greater risk, whereas jobs requiring social, emotional and literary abilities are at the lowest risk of displacement.

A positive future with artificial intelligence to bring remote online office working environment chance:

Many businesses and individuals are optimistic that this AI-driven shift in the workplace will result in more jobs being created than lost. As we develop innovative technologies, AI will have a positive impact on our economy by creating jobs that require the skill set to implement new systems. 80% of respondents in the EY survey said it was the lack of these skills that was the biggest challenge when employing AI programs.

It is likely that artificial intelligence will soon replace jobs involving repetitive or basic problem-solving tasks, and even go beyond current human capability. AI systems will be making decisions instead of humans in industrial settings, customer service roles and within financial institutions. Automated decisioning will be responsible for tasks such as approving loans, deciding whether a customer should be onboarded or identifying corruption and financial crime.

Organisations will benefit from an increase in productivity as a result of greater automation, meaning more revenue will generated. This thus provides additional money to spend on supporting jobs in the services sector.Due to the vast array of jobs that could be impacted by AI, it is fundamental to address the potential pitfalls of these technologies. Business need to overcome the trust and bias issues surrounding AI by achieving an effective and successful implementation that makes it possible for everyone to benefit.

Governments must ensure that gains from AI are shared widely across society to prevent social inequality between those affected and unaffected by these developments. For example, this could be through increased investment into training. With the additional cost-savings from implementing AI systems, employers should also focus on upskilling their current employees.

To properly leverage the power of AI, we need to address the issue at an educational level, as well as in business. Education systems needs to focus on training students in roles directly associated to working with AI, including programmers and data analysts. This requires more emphasis to be put on STEM subjects (science, technology, engineering and mathematics). Also, subjects centered around building creative, social and emotional skills should be encouraged. Whilst artificial intelligence will be more productive than human workers for repetitive tasks, humans will always outperform machines in jobs requiring relationship-building and imagination. Hence, artificial intelligence will change our world both inside and outside the workplace. Instead of focusing on the fear surrounding automation, businesses need to embrace these new technologies to ensure they implement the most effective AI systems to enhance and compliment human intelligence.

Artificial Intelligence (AI) in Banking working environment

Artificial Intelligence (AI) is a fast-evolving technology, gaining popularity all around the world. Several industries have already adopted AI for various applications, getting better and smarter day by day. In the past few years, the banking sector has also become one of the leading adopters of Artificial Intelligence. Most banks and financial institutions are implementing AI to add more efficiency to their back-office and lessen security risks.

As per Statista, the AI market in the United States is forecasted to reach 7.35 billion U.S. dollars in 2018. Some major applications of AI include classification, image recognition, object identification, and automated geophysical feature detection. Speaking of banking and financial institutions, JPMorgan Chase, Wells Fargo, Bank of America, CitiBank, and other leading U.S. banks have already implemented AI in their systems, helping consumers manage their daily banking needs more efficiently.

AI technology can bring better Customer Support in bank service environment

Several pieces of evidence advocate that the customers willingly prefer self-service options which allow them to chat with a virtual assistant as if it were a live customer representative. Most leading banks have already added virtual assistants to their instant website chatbots, voice response systems, and mobile applications. Artificial Intelligence considers each interaction as a teachable moment, so the chatbots (virtual assistants) keeps getting better while understanding customers. With AI, virtual assistants can deliver better customer support. It also allows sentiment analysis, so the virtual assistant can determine when individuals are getting frustrated and instantly transfer them to a live agent.

Enhanced Banking Services

AI streamlines the banking process while giving customer service a new level of comfortability. It allows banks to meet customers' expectations with comprehensive digital support. With Artificial Intelligence, you can achieve greater precision and accuracy. From cash transfer to bills payment, cards management, and other support, AI can significantly enrich the satisfaction level of your customers. All of these operations can be easily managed through desktops, smartphones, and other mobile devices.

Scam Recognition

With an immense growth of banking fraud, scam recognition and reduction has become challenging for the banking sector. Several banks tried to identify the factors and powerful solutions but couldn't succeed. However, AI makes it easier to detect the factors involved in frauds and support investigators. It improves financial security with advanced fraud prevention tactics. Artificial Intelligence works as a real-time scam solution for the banking sector while handling complex situations and tactics. Based on advanced data crunching, AI can detect fraud by flagging unusual transactions. It also feeds back into the consumer's profile which subsequently builds a secure environment.

Advanced Data Analytics

One of the main advantages of AI is its ability to complete tedious tasks through intricate automation, resulting in better productivity. Based on a machine learning algorithm, AI can quickly consume and process a massive amount of data at an expedited level. The enormous speed brings efficiency to financial services, providing scope for personalized offerings to consumers. What's even more, AI makes faster decisions while carrying out actions quickly. With such advantages, it is nearly obvious that the majority of banks and financial institutions will adopt AI to stay competitive and deliver better customer support. However, several cons are also associated with a machine learning algorithm. As it continues to learn and grow, the decision-making capabilities may create problems in the near future.

Disadvanages of AI in Banking Sector

Artificial intelligence is also expected to massively disrupt banks and traditional financial services. Some of its disadvantages are listed below.

Highly Expensive

Production and maintenance of artificial intelligence demand huge costs since they are very complex machines. AI also consists of advanced software programs which require regular updates to meet the needs of the changing environment. In the case of critical failures, the procedure to reinstate the system and recover lost codes may require enormous time and cost.

Bad Calls

Though Artificial Intelligence can learn and improve, it still can't make judgment calls. Humans can take individual circumstances and judgment calls into account when making decisions, something that AI might never be able to do. Replacing adaptive human behavior with AI may cause irrational behavior within ecosystems of humans and things.

Distribution of Power

There is a constant fear of AI superseding or taking over the humans.

Artificial intelligence can give a lot of power to the few individuals who are controlling it. Hence, AI carries the risk and takes control away from humans while dehumanizing actions in several ways.

Unemployment

Replacement of the workforce with machines can lead to wide-reaching unemployment. Moreover, if the use of AI becomes rampant, people will be highly dependent on the machines and lose their creative power. Unemployment is a socially undesirable issue. Individuals with nothing to do can lead to the devastating use of their minds. Be it banking or any other sector; Artificial intelligence can effectively increase the unemployment rate.

Artificial Intelligence delivered to wrong hands can turn out to be a serious threat to humankind. If individuals start thinking destructively, they can generate havoc with these advanced machines. The challenges introduced by the emergence of artificial intelligence revolve around several things. However, AI is a right balance of skill and emotions which is continually growing. Artificial intelligence provides banks, financial institutions, and tech companies with significant competitive advantages. Nevertheless, it can completely transform the financial sector and make it faster, but this will only be possible if the financial industry can manage the security risk of systems based on AI.

What does artificial intelligence mean for the bank service office workers?

With all these new artificial intelligence use cases comes the question of whether machines will force humans into obsolescence. The jury is still out: Some experts vehemently deny that artificial intelligence will automate so many jobs that millions of people find themselves unemployed, while other experts see it as a pressing problem.

"The structure of the workforce is changing, but I don't think artificial intelligence is essentially replacing jobs in bank service working environment. It allows us to really create a knowledge-based economy and leverage that to create better automation for a better form of life. It might be a little bit theoretical, but I think if you have to worry about artificial intelligence and robots replacing some bank service jobs, e.g. bank security, bank enquiry service,. But, AI can not replace bank counter service staffs to do saving or withdrawing money transfer tasks when any customers prepare to save money or withdraw money in bank counters. As this technology develops, the AI bank service will see new startups, numerous saving or withdraw transactions from consumer won't be raise more easily.

AI to Banking and Finance industry

The banking and finance industry plays a major role in our lives. I mean the world runs on money and banks are essentially the gatekeepers that regulate that flow. Did you know that the banking and finance industry heavily relies on artificial intelligence for things like customer service, fraud protection, investment, and more? A simple example is the automated emails that you receive from banks whenever you do an out of the ordinary transaction. Well, that's AI watching over your account and trying to warn you of any fraud.

AI is also being trained to look at large samples of fraud data and find a pattern so that you can be warned before it happens to you. Also, when you hitch a little snag and chat with bank's customer service, chances are that you are chatting with an AI bot. Even the big players in the finance industry use AI to analyze data to find the best avenues to invest money so they can get the most returns with the least risk. That's not all, AI is poised to play an even bigger role in the industry as major banks across the world are investing billions of dollars in the AI technology and we all will observe its effects sooner than later

How AI influences our daily working life in any office working environment

Can AI bring only disadvantages? If AI can bring disadvantges, what are its disadvantages to any working environment ?The entire tech world is debating the consequences of artificial intelligence and the part AI is going to play in shaping our future. While we might think that artificial intelligence is at least a few years away from causing any considerable effects on our lives, the fact remains that it is already having an enormous impact on us. Artificial intelligence is affecting our decisions and our lifestyles every day. Don't believe me? I shall indicate some product examples how AI anticipates which can influence our working culture in any office environment.

Examples of how Artificial Intelligence assistance to office working environment may include as below:

1. Smartphones

Smartphones have become the most indispensable tech product that we own today and we use it almost all the time. Well, if you are using a smartphone, you are interacting with AI whether you know it or not. From the obvious AI features such as the built-in smart assistants to not so obvious ones such as the portrait mode in the camera, AI is impacting our lives in every day office working environment.

In fact, the two examples that I provided that our working world of AI and how it is effecting our working lives. Firstly, there are the obvious AI elements which most of us have some knowledge about. For example, when you are using a smart assistant in office, whether it's Google Assistant, Alexa, Siri, or Bixby, you more or less know that these assistants are based on AI. However, when we are using a feature such as the portrait mode effect while shooting a picture, we never consider that AI might be behind that too. Have you ever thought how the Google Pixel phones or iPhones can capture such great portrait shots? The answer is artificial intelligence. So, when any office workers need to find any knowledge to solve their working problem immediately in any offices. They may apply AI smart phone tools to help them to apply online channel to search any new knowledge to attempt to solve their working problem in possible, when their computers have none any computers in offices.

Now more and more manufacturers are including AI in their smartphones with big chip manufacturers including Qualcomm and Huawei producing chips with built-in AI capabilities. The AI integration is helping in bringing features like scene detection, mixed and virtual reality elements, and more. AI is going to play an even major role in the coming years. We are already seeing the huge emphasis on AI with the latest Android and iOS updates. Features like app actions, splices, and adaptive battery in Android Pie and Siri shortcut and Siri suggestions in iOS 12 are made possible with AI. So, next time if any office workers think AI is not effecting them, take out your smartphone to replace computers to find any knowledge to help you to solve any tasks problems immediately in offices.

2. Social Media Feeds

If you are thinking that smart cars don't personally effect you as they are still not in your country or city, well, how about something which you use on a daily basis. Even if you are living under a rock, there's a high probability that you are tweeting from underneath it. If Twitter's not your choice of poison, maybe it's Facebook or Instagram, or Snapchat or any of the myriad of social media apps out there. Well, if you are using social media, most of your decisions are being impacted by artificial intelligence. So, any office workers may apply AI to help them to gather any new information to solve any difficult task problems , if their managers can not assist them to solve any sudden tasks problem, they are encountering to need to solve any working complex tasks problem internet social media in any any office working environment immediately.

From the feeds that office staffs can see in their working timeline to the notifications that you receive from these apps, everything is curated by AI. AI takes all your past behavior, web searches, interactions, and everything else that you do when you are on these websites and tailors the experience just for you. The sole purpose of AI here is to make the apps so addictive that you come back to them again and again, and I am ready to place a bet that AI is winning this war against you.

3. Online Ads Network

One of the biggest users of artificial intelligence is the online ad industry which uses AI to not only track user statistics but also serve us ads based on those statistics. Without AI, the online ad industry will just fail as it would show random ads to users with no connection to their preferences what so ever. AI has become so successful in determining our interests and serving us ads that the global digital ad industry has crossed 250 billion US dollars with the industry projected to cross the 300 billion mark in 2019. So next time when any product developers are going online and seeing ads or product recommendation, know that AI is impacting to any new products advertisement method more efficiently.

4. AI can be any office security

While we can all debate the ethics of using a broad surveillance system, there's no denying the fact that it is being used and AI is playing a big part in that. It is not possible for humans to keep monitoring multiple monitors with feeds from hundreds if not thousands of cameras at the same time, and hence, using AI makes perfect sense. With technologies like object recognition and facial recognition getting better and better every day, it won't be long when all the security camera feeds are being monitored by an AI and not a human. While there's still time before AI can be fully implemented such as security in any offices, this is going to be our future.

5. Smart Keyboard Apps

Smart Keyboard Apps. Granted, not everyone loves dealing with on-screen keyboards. However, they have become far more intuitive, allowing users to type comfortably and faster. What has probably proved to be a catalyst for them is the integration of AI. The smart keyboard apps keep a tab on the writing style of a user and predict words and emojis accordingly. Thus, typing on the touchscreen has become faster and more convenient. Not to mention, artificial intelligence also plays a vital role in pin-pointing misspellings and typos. So, any office workers can apply smart keyboard apps to help their to raise typing efficiency and reduce wrong typing word in error when they need to type any document in offices.

6. E-Commerce

` AI-driven algorithms have kind of given the much-needed impetus to e-commerce to provide a more personalized experience. According to several reports, its usage has vastly increased sales and also played a good part in building loyal relationships with customers. Thus, companies take advantage of AI to deploy chatbots to collect pivotal data and also predict purchases to create a customer-centric experience. Yet to come across this shift of strategy? Just spend some time with sites like Amazon and eBay and you will soon get to know how fast the landscape is changing around you – for the better! So, Ai can help any businesses to achieve e-commerce sale channel more easily.

7. Smart Email Apps

In any office working environment, if you still find your inbox cluttered with too many unwanted messages, chances are pretty high that you are still stuck with an old school email app. You heard it right! Modern email apps like Spark make the most of AI to get rid of spam messages and also categorize emails so that you can quickly access the important ones. What's more, they also offer smart replies based on the messages you receive to help you reply to any email quickly. The "Smart Reply" feature of Gmail is a great example of this. It uses AI to scan the text of the email and provides you with contextual answers. So, AI can help any office staffs to know who had sent any message from email and respond their email immediate , when AI can help any offices to avoid to receive any email spam rubbish email message in any time, even after working hours, it means that AI is working to help any office staffs to avoid to receive any email spam rubblish message in any time. So, when they go to office to work, even they go home after working hours. They can know whether what the important email messages are sent to their office email in boxes any time. Then, they can send email to respond their customers' enquires any time. So, AI can help any office workers can have chance to work at homes.

The Future of Artificial Intelligence In The Workplace

Smart technologies aren't just changing our homes; they're edging their way into their numerous industries and are disrupting the workplace. Artificial Intelligence (AI) has the potential to improve productivity, efficiency and accuracy across an organization – but is this entirely beneficial? Many fear that the rise of AI will lead to machines and robots replacing human workers and view this progression in technology as threat rather than a tool to better ourselves.

With AI continuing to be a prominent buzzword in 2019, businesses need to realize that self-learning and black-box capabilities are not the panacea. Many organisations are already beginning to see the incredible capabilities of AI, using these advantages to enhance human intelligence and gain real value from their data. As there is increasing evidence demonstrating the benefits of intelligent systems, more decision-makers in the boardroom are gaining a better understanding of what AI can really offer. Research conducted by EY explains "organizations enabling AI at the enterprise level are increasing operational efficiency, making faster, more informed decisions and innovating new products and services."

Today In: Cybersecurity

The first companies employing AI systems across the board will gain competitive advantage, reduce cost of operations and remove head counts. Whilst this may be a positive from a business perspective, it is obvious why this a worry for those working in roles at risk of displacement. The introduction of these technologies will likely trigger an issue with unions and job security due to the substantial operational changes. Although AI will affect every sector in some way, not every job is at equal risk. PwC predicts a relatively low displacement of jobs (around 3%) in the first wave of automation, but this could dramatically increase up to 30% by the mid-2030's. Occupations within the transport industry could potentially be at much greater risk, whereas jobs requiring social, emotional and literary abilities are at the lowest risk of displacement.

A positive future with artificial intelligence

Many businesses and individuals are optimistic that this AI-driven shift in the workplace will result in more jobs being created than lost. As we develop innovative technologies, AI will have a positive impact on our economy by creating jobs that require the skill set to implement new systems. 80% of respondents in the EY survey said it was the lack of these skills that was the biggest challenge when employing AI programs. It is likely that artificial intelligence will soon replace jobs involving repetitive or basic problem-solving tasks, and even go beyond current human capability. AI systems will be making decisions instead of humans in industrial settings, customer service roles and within financial institutions.

Automated decisioning will be responsible for tasks such as approving loans, deciding whether a customer should be onboarded or identifying corruption and financial crime. Organisations will benefit from an increase in productivity as a result of greater automation, meaning more revenue will generated. This thus provides additional money to spend on supporting jobs in the services sector.

How to take advantage of AI to any offices

Due to the vast array of jobs that could be impacted by AI, it is fundamental to address the potential pitfalls of these technologies. Business need to overcome the trust and bias issues surrounding AI by achieving an effective and successful implementation that makes it possible for everyone to benefit. Governments must ensure that gains from AI are shared widely across society to prevent social inequality between those affected and unaffected by these developments. For example, this could be through increased investment into training.With the additional cost-savings from implementing AI systems, employers should also focus on upskilling their current employees.

To properly leverage the power of AI, we need to address the issue at an educational level, as well as in business. Education systems needs to focus on training students in roles directly associated to working with AI, including programmers and data analysts. This requires more emphasis to be put on STEM subjects (science, technology, engineering and mathematics). Also, subjects centered around building creative, social and emotional skills should be encouraged. Whilst artificial intelligence will be more productive than human workers for repetitive tasks, humans will always outperform machines in jobs requiring relationship-building and imagination. Artificial intelligence will change our world both inside and outside the workplace. Instead of focusing on the fear surrounding automation, businesses need to embrace these new technologies to ensure they implement the most effective AI systems to enhance and compliment human intelligence

How AI can help office workers to do tasks more easily

Companies are currently spending big on artificial intelligence and machine learning initiatives to the tune of $12 billion, but estimates put that figure as high as $57.6 billion by 2021, according to the International Data Corporation (IDC). With such massive shifts, the focus is usually on what we might lose, but it shouldn't be. A recent report on the future of work from the McKinsey Global Institute suggests that while only about 5% of jobs can be completely eliminated by automation, the rise of AI requires workers to beef up both technical and soft skills in order to stay competitive.

What's seldom discussed is how AI can revolutionize our jobs. It's now possible to pinpoint peak productivity for a single day, improve communication in meetings (even before people ever work together face to face), or even teach you to be a better leader, all thanks to AI platforms. I shall indicate these advantages to bring any office benefits from AI assistance as below:

1. AI can help any companies to get better to hire the best applicants

AI has the greatest potential to change the way companies find candidates, according to Alexander Rinke, cofounder and CEO of Celonis. The company's process-mining technology helps businesses to understand the areas where automation can help humans, he says. In HR departments, Celonis can help identify how fast workers come and go, the cost per hire, and which positions take the longest to fill. AI helped enable one customer's ability to identify bottlenecks in recruitment and reduced process costs internally by 30% as well as get them hired more quickly, he says.

Crafting a resume has never been easier, nor has landing an interview. Another example is how recruitment software provider iCIMS, in partnership with Google, is helping job seekers find jobs directly through the search engine, thanks to Google's AI and machine learning capabilities. Susan Vitale, iCIMS's chief marketing officer says that in addition to reducing the number of expired job postings, machine learning is underlying a private beta program of Google's Cloud Jobs Discovery model. "For a candidate searching for, say, a CTO role, Cloud Job Discovery will serve up CTO positions as well as jobs with titles that are similar, but not verbatim, such as chief technology officer or chief technical officer," says Vitale. This model also allows for conceptual search results, such as serving up job listings for cashiers, sales associates, and store associates when someone searches for one versus just only showing jobs that exactly match the keyword search criteria, she adds.

2. AI can help any office workers to raise much more productive efficiencies

John Furneaux, CEO and cofounder of Hive, says predictive analytics will help us better understand how we work. "It can tell us just about everything we want to know about teams and collaboration, for example, if men or women get more done in the afternoon, and if summer Fridays are a myth," he says. (Everyone thinks summer Fridays aren't productive, but in reality there's no difference between those and other Fridays during the year–productivity is equally low.)

Using a data set of over 30,000 completed actions across Hive workspaces, Furneaux says they were able to identify some notable trends in productivity. For example, men were far more productive early in the day, with a sharp decline in the afternoon, while women had a slower start to the day but were far more productive in later hours than their male counterparts. And analyzing chat messages revealed that women appear to complete more tasks when chatting, suggesting they use communication as a key tool to completing work. Similarly, Nintex Hawkeye analyzes data on business processes by types, users, roles, and departments to see who's doing the work and how long it takes them to do it. Management can monitor and analyze those metrics in real time.

3. AI can help any managers to make the most fair compensation and eliminate wage gaps to every staffs

Tanya Jansen, cofounder of the compensation management platform beqom, says that AI and predictive analytics can eliminate unconscious bias from compensation. Jansen says that AI based on a variety of rules including education, experience, certifications, and more can make compensation more fair and help businesses move closer to closing pay gaps. "Specifically, AI can help solve gender pay gaps and the CEO-to-worker pay gap, in which pay ratios of Fortune 500 companies range from 2:1 at the low end to nearly 5000:1 at the high end," she says. Additionally, the use of AI-driven compensation technology to make pay more fair can mitigate the risk of employee turnover, which costs businesses as much as 33% of a worker's annual salary to replace them.

4. AI can help any office staffs to arrange better meetings

Augmented Reality (AR) is still in its infancy, but AI and machine learning are the core components that make it work. As such, Christa Manning, the vice president and solution provider research leader at Bersin, Deloitte Consulting LLP, says that AR can help workers find the right information, in the right place, at the right time to make the best decisions wherever they may be working. For example, as more companies adopt video meetings and collaborative workspaces, it's likely we'll begin to see HR-curated information like talent profiles and work styles layered over interactions through AR."Imagine being in a video conference with a colleague and having direct insight into their communication style, seeing tips on how to best interact with them or reminders of what needs to be discussed. SO, AI can help any organizations to conclude or find the best methods to solve any problems after their every discussion in any meetings.

How AI is improving onboarding and training. AI coaching tools first learn by observing how different employees conduct specific tasks. Then these tools can walk new employees through how to complete those tasks—or even coach existing employees on how to do things more effectively or efficiently. Chorus is a great example of this technology. It analyzes sales calls while they happen, offering tips to help sales reps manage the cadence of meetings and use the most effective messaging. It also records all sales calls and compiles statistics for each sales rep, providing everyone with the tools they need to help them close more deals and conduct more effective calls. Another example is Cogito, a tool that combines AI with behavioral science to help customer service employees provide better phone support. It monitors calls for voice signals, providing real-time suggestions to representatives on how to improve the conversation.

5. AI can help any managers to be better leaders

Indiggo, a platform powered by a proprietary AI tool called "indi," functions as a brain that has consumed all the knowledge the company has gathered in its 15 years of operation. It also uses an algorithm to provide an estimate of how much time is wasted by a company by analyzing the size of its management team. Then it taps their calendars to see how they spend their time, and walks individual managers through a type of Q&A to make sure they are clear on what their top three priorities are, and how that relates to the organization's priorities, which will indicate if that strategy is moving forward or not. "The counterintuitive impact of these advances is that they actually make human work truly irreplaceable," Alexander Rinke, the cofounder and CEO of Celonis says. As such, he reminds us, "Humans are much better at processes that involve reasoning, judgment, and interaction with people." So, AI can recommend more accurate and useful opinions to help any managers to solve their managing challenges in office any time.

How AI is eliminating repetitive administrative tasks

There are a lot of tasks that knowledge workers spend time on that provide little—if any—value.For example, say you need to schedule a meeting to get consensus on a decision before moving forward, but you need five people to join the meeting. It's easy to spend a ton of time sending email back-and-forth or finding an open slot on everyone's calendar.That's not the most rewarding use of your time for you or your company.Tools like X.ai give employees AI-powered personal assistants that perform administrative tasks like scheduling, rescheduling, and cancelling meetings.

How AI is transforming internal communications and support

Personnel on the teams that provide employee support have their hands full with other responsibilities, too. HR teams work on building the kind of company people love working for. IT maintains the company's network and keeps data secure. Office managers frequently run big events like holiday parties.These tasks are crucial, but they're often hard for teams to focus on because they're busy answering routine questions. AI service desks like askSpoke allow employee support teams to balance their service commitments with other important responsibilities by reducing interruptions from rote, repetitive requests.Employees can askSpoke for whatever they need over Slack, email, SMS, and the web. askSpoke's friendly AI will automatically provide a prompt response.

How AI is transforming marketing, sales, and customer service

AI-powered chatbots help with external support as well. Just like with internal support tools like askSpoke, these chatbots learn from real marketers, salespeople, and customer service reps and are eventually able to answer questions as accurately as a knowledgeable person.For example, chatbot for Messenger helps customers plan their vacations. It books flights, hotels, and cars, highlights destination attractions, and even provides answers to questions like "Where can I go for $100 expense budget only?"

How AI is transforming business data and analytics

It's hard to run a competitive business today without data. But even massive amounts of data are useless without a way to transform that data into valuable insights. That's typically why you'd want to hire a data scientist—which just happens to be one of the most difficult roles to fill. How AI is fighting fraud and transforming security. Have you ever taken a call from your bank to find that someone used your debit card fraudulently? Most likely, your bank used some form of AI to detect the fraudulent transaction and decline it. Applying the same basic technology to the workplace helps identify security risks and keeps customer, employee, and company data safe. AI-powered software can automatically detect and address threats among thousands or millions of signals that humans would never be able to parse (especially not in real-time).

How AI is transforming productivity

While AI is transforming the workplace in many different ways across every industry, it's impacting productivity most of all. When your office staffs don't have to scroll through calendars to look for open meeting times, build reports in spreadsheets to look for insights, or spend your day answering the same questions over and over again, you're more productive. Workers are freed from redundant and mindless tasks, giving them more time to do work that matters, solve problems, and exercise their creativity. Some tools use AI to specifically monitor and boost productivity. For example, Deloitte's LaborWise provides company leaders and managers with productivity analytics that help them identify areas where labor costs are too high, impediments that slow people down, and departments that need additional staff.

In conclusion, what AI means for the workplace of the future. While some will dramatize the negative impacts of AI, cognitive computing, and robotics, these powerful tools will also help create new jobs, boost productivity, and allow workers to focus on the human aspects of work. Essentially, automation frees companies and their employees up to be more empathetic, to focus on things like the customer experience, employee engagement, and workplace culture.

What are traditional office tools to be replaced by AI ?

Artificial intelligence (AI) is predicted to eliminate over a million jobs in the next few years, potentially replacing lower level positions like administrative assistants with humanoid robots or voice assistants. But in the nearer future, fresh AI-driven software and products are also moving to eliminate non-human elements of the workplace by replacing traditional office tools, including both physical products and everyday electronic processes. Why should businesses switch from the tried-and-true to emerging technology? Many of the experts TechRepublic talked to said the AI options streamline business practices, making their adopters work smarter instead of harder. I shall indicate these office tools ,they can be applied to help any office staffs to finish their these tasks in office, they may include as below:

1. Scheduling

Workloud's end-to-end, cloud-based workforce management software takes scheduling from paper or Excel and moves it to the cloud. Everything from clocking in and out to monitoring employee absences is fully digitalized.Schedules and timesheets are accurate, created easily, and accessible through the service's web, tablet, and mobile apps. The software can also be used for absence management.

2. Employee talent selection

Using AI and organizational behavior science, can be used to replace internal spreadsheets and databases designed to monitor human capital. By mining employee attributes and experiences, the software can recommend who would be best for a project. The software also collects reviews after projects to better predict successful employee-project matches.The traditional hiring process is slow, biased and inaccurate, By removing humans from the beginning stages of the process, it can become faster and more fair, and result in better hires.

AI software automates the hiring process, using online simulations instead of manual screenings and interviews. Using the software, employers can include tasks in a job application, allowing job candidates to show technical skills that may be necessary for a job. Employers can't rule out candidates until they see how the candidate performs, eliminating bias that occurs in the resume reading stage. Both sides also automatically receive updates about each other's steps, reducing the amount of time it takes to .

3.Timesheets: Allocate

Using AI and machine learning, the software registers an employee's computer activity throughout the day. The data, which can also pull information from email and calendars, is used to suggest timesheet entries to reflect a more accurate amount of time an employee spent working. The employee can review and revise as necessary. However, the software doesn't spy on or monitor employees. The data is only available to each employee, while others in the company can only see the timesheet's output, which Allocate said would be the same information available if a manual sheet was used. So,replacing manual timesheets with Allocate has three advantages: More accurate time entry, project analytics, and "'unsucking' the work experience."

4. Document storage

By using AI to read and analyze business and legal documents, AI can store all of the important document-based information in the cloud. The severe reduction in print-outs means less paper and ink, fewer products like binder clips and boxes to store and organize all of the paper, and more employee time freed up from not needing to manually sort through every document.

For example, in any lawyer offices, legal professionals' morale in the industry can suffer when they are pushed into performing such dull, repetitive tasks like sorting through and coding documents by hand, With AI tools to automate those duties, lawyers can focus on more meaningful projects and boost the business's and clients' success as a result. While focused on law firms, businesses that have a lot of unstructured data in documents may also be able to use the service to free up employee time and save on printing costs.

5. Scanners: Adobe Scan

While documents are moving to the cloud more and more, sometimes a physical copy of a document still needs to be scanned using a bulky office scanner. Adobe Scan, an app that condenses a scanner to the size of a smartphone, can rid offices of the need for an in-house scanner. Users can download and open the app, then hold their device over whatever they need to scan. Adobe Sensei then turns the scan into a PDF, and sends it to the Adobe Document Cloud. The app can transform any image into digital text that can then be searched and used electronically. The app streamlines the scanning process, making scans cleaner and more immediate. For businesses already using Adobe services, the app makes documents easily accessible.

6. Landline phones

While landlines in homes are increasingly less common, the same cannot be said for offices. But using chatbots and AI integrations, RingCentral is trying to replace traditional office landline phone systems. The platform offers over 100 integrations, including that AI landline phones can let employees check their voicemail, and a Gong.io option that listens to call recordings to find traits of successful employees than can be used in training. An add-on for Gmail lets users switch from emailing back and forth to a voice session without needing to look up contact information. AI landline phone is easy to adopt and use in the workplace, and is more customizable than standard phone systems, said David Lee, vice president of platform products. Compared to the traditional option, the cloud-based option is "future-proof.

How artificial intelligence can raise office efficiency

Artificial Intelligence is already impacting every industry through automation and machine learning, bringing concerns that AI is on the fast track to replacing many jobs. But these fears aren't new, says Dan Jackson, director of Enterprise Technology at Crestron, a company that designs workplace technology. "I'd argue this is no different than when we moved from an agricultural to an industrial economy at the turn of the last century. The percentage of people working in agriculture significantly decreased, and it was a big shift, but we still have plenty of jobs 100 years later," he says. Anytime society experiences a major technological advancement, we need to be prepared for it to change the way we live and work. It's hard to imagine what the future of jobs will look like with AI, but that future exists. And optimists suggest that, like the sewing machine to the textile industry, AI will make us better, more efficient and faster workers.

In fact, many experts agree that AI has the potential to eliminate mundane, administrative work, while we will always rely on human workers to be empathetic, collaborative, creative and strategic. But it's impact on any industry lies in the hands of the business leaders who are responsible for adopting AI strategies.

● Training presents challenges

A recent study of 1,000 global companies by Accenture found that AI is already creating three new categories of jobs: trainers, explainers and sustainers. Trainers are the people who teach AI systems how to act -- whether it's language, human behavior or the intricacies of human interaction. Explainers are the liaison between technology and business leaders, providing more insight and clarity into machine learning for the non-tech workers. Sustainers are the workers required to maintain AI systems and troubleshoot any potential issues. Some jobs were highly technical and required advanced degrees, but other roles demanded innately human things such as empathy and interaction. Downstream jobs, such as those in sales, marketing, or service will change to take advantage of the insights from AI, but many of the core skills will remain. However, it might sound like any job related to AI will require years of technical knowledge, but that isn't the case. We've already seen a shift in tech hiring -- companies often need highly specific skill sets that are hard to find in potential candidates. As a result, more businesses are hiring employees with the right soft skills, and then training them in technical skills.

An office effort measured approach to AI

The real takeaway is that any approach to AI will need to consider the human aspect of every business. AI has great potential to increase efficiency and accuracy and it's already been proven in certain industries. For example, the use of AI In banking to identify and money laundering schemes. It's also improved healthcare by "increasing the speed and accuracy" of cancer diagnosistics. AI can also help reduce the cost and length of human trafficking investigations, a situation where time is precious. In these examples, AI hasn't replaced jobs, but has positively impacted efficiency.

Thus, we need to ensure our education system responds to equip young people with the appropriate skills and adaptability, while businesses and public organizations must invest in training. Perhaps most of all, we need to encourage imagination and willingness to experiment. The organizations that can innovate with AI will reap the benefits. Their growth will make them the primary source of future jobs. Companies have a choice when implementing AI. They can choose to effectively implement systems that make employee's lives easier and find creative ways to leverage the technology. It's up to employers to ease fears for workers around AI and build strategies that benefit everyone. Hence, some AI experts believe AI can only raise efficiency to some office tasks, however, AI can not still raise efficiency to all office tasks for any office deparments. The reasons are because some office tasks which can only dominate to finish by human office workers. These office tasks are as below:

How can leaders and managers improve employee productivity while still saving time? These below tasks, AI experts ensure that AI can not help any office workers to raise their efficiencies as below:

1. Office managers can not delegate to AI to help them to do. While this tip might seem the most obvious, it is often the most difficult to put into practice. We get it–your company is your baby, so you want to have a direct hand in everything that goes on with it. While there is nothing wrong with prioritizing quality (it is what makes a business successful, after all), checking over every small detail yourself rather than delegating can waste everyone's valuable time. Instead, give responsibilities to qualified employees, and trust that they will perform the tasks well. This gives your employees the opportunity to gain skills and leadership experience that will ultimately benefit your company. You hired them for a reason, now give them a chance to prove you right.

2. Office managers can not match Tasks to Skills to AI. Knowing your employees' skills and behavioral styles is essential for maximizing efficiency. For example, an extroverted, creative, out-of-the-box thinker is probably a great person to pitch ideas to clients. However, they might struggle if they are given a more rule-intensive, detail-oriented task. Asking your employees to be great at everything just isn't efficient–instead, before giving an employee an assignment, ask yourself: is this the person best suited to perform this task? If not, find someone else whose skills and styles match your needs.

3. Office managers can not teach AI to replace them how to communicate and teach their low level staffs how to work effectively. Every

manager knows that communication is the key to a productive workforce. Technology has allowed us to contact each other with the mere click of a button (or should we say, tap of a touch screen)–this naturally means that current communication methods are as efficient as possible, right? Not necessarily. A McKinsey study found that emails can take up nearly 28% of an employee's time. In fact, email was revealed to be the second most time-consuming activity for workers (after their job-specific tasks). Instead of relying solely on email, try social networking tools (such as Slack) designed for even quicker team communication. You can also encourage your employees to occasionally adopt a more antiquated form of contact...voice-to-voice communication. Having a quick meeting or phone call can settle a matter that might have taken hours of back-and-forth emails. All of above communication tasks, I believe that AI can not do better than managers in offices.

4. AI can not keep Goals Clear and focused to be better than managers. You can't expect employees to be efficient if they don't have a focused goal to aim for. If a goal is not clearly defined and actually achievable, employees will be less productive. So, try to make sure employees' assignments are as clear and narrow as possible. Let them know exactly what you expect of them, and tell them specifically what impact this assignment will have. One way to do this is to make sure your goals are "SMART" – specific, measurable, attainable, realistic, and timely. Before assigning an employee a task, ask yourself if it fits each of these requirements. If not, ask yourself how the task can be tweaked to help your workers stay focused and efficient.

5. AI can not know how to incentivize Employees to work more efficiently. One of the best ways to encourage employees to be more efficient is to actually give them a reason to do so. Recognizing your workers for a job well done will make them feel appreciated and encourage them to continue increasing their productivity. When deciding how to reward efficient employees, make sure you take into account their individual needs or preferences. For example, one employee might appreciate public recognition, while another would prefer a private "thank you." In addition to simple words of gratitude, here are a few incentives managers can know how to incentivize their staffs to work efficiently, but AI is only one machine, it can not perform very good.

6. AI does not know how to assist managers to train and Develop employees. Reducing training, or cutting it all together, might seem like a good way to save company time and money (learning on the job is said to be an effective way to train, after all). However, this could ultimately backfire. Forcing employees to learn their jobs on the fly can be extremely inefficient.

So, instead of having workers haphazardly trying to accomplish a task with zero guidance, take the extra day to teach them the necessary skills to do their job. This way, they can set about accomplishing their tasks on their own, and your time won't be wasted down the road answering simple questions or correcting errors. Past their original training, encourage continued employee development. Helping them expand their skillsets will build a much more advanced workforce, which will benefit your company in the long run. There are a number of ways you can support employee development: individual coaching, workshops, courses, seminars, shadowing or mentoring, or even just increasing their responsibilities. Offering these opportunities will give employees additional skills that allow them to improve their efficiency and productivity. But, AI do not know how to improve any office workers' performance more easily than managers.

● How can AI be dangerous to office working environment?

Most researchers agree that a superintelligent AI is unlikely to exhibit human emotions like love or hate, and that there is no reason to expect AI to become intentionally benevolent or malevolent. Instead, when considering how AI might become a risk to any office working environments, experts think two scenarios most likely:

The AI is programmed to do something devastating: Autonomous weapons are artificial intelligence systems that are programmed to kill. In the hands of the wrong person, these weapons could easily cause mass casualties. Moreover, an AI arms race could inadvertently lead to an AI war that also results in mass casualties. To avoid being thwarted by the enemy, these weapons would be designed to be extremely difficult to simply "turn off," so humans could plausibly lose control of such a situation. This risk is one that's present even with narrow AI, but grows as levels of AI intelligence and autonomy increase. So, if some businessmen apply AI to be business weapon to attack or steal their business competitors' business secret, e.g. contract document, employee performance report, profit report, even business secret document. Then, AI will be one business competitor weapon more than business assistant role in any business market. So, whether AI is office assistant or business competitor weapon, it depends on how the businessmen apply them to assist their business development.

The AI is programmed to do something beneficial, but it develops a destructive method for achieving its goal: This can happen whenever we fail to fully align the AI's goals with ours, which is strikingly difficult. If you ask an obedient intelligent car to take you to the airport as fast as possible, it might get you there chased by helicopters and covered in vomit, doing not what you wanted but literally what you asked for. If a superintelligent system is tasked with a ambitious geoengineering project, it might wreak havoc with our ecosystem as a side effect, and view human attempts to stop it as a threat to be met.

As these examples illustrate, the concern about advanced AI isn't malevolence but competence. A super-intelligent AI will be extremely good at accomplishing its goals, and if those goals aren't aligned with ours, we have a problem. You're probably not an evil ant-hater who steps on ants out of malice, but if you're in charge of a hydroelectric green energy project and there's an anthill in the region to be flooded, too bad for the ants. A key goal of AI safety research is to never place humanity in the position of those ants. Because AI has the potential to become more intelligent than any human, we have no surefire way of predicting how it will behave. We can't use past technological developments as much of a basis because we've never created anything that has the ability to, wittingly or unwittingly, outsmart us. The best example of what we could face may be our own evolution. People now control the planet, not because we're the strongest, fastest or biggest, but because we're the smartest. If we're no longer the smartest, are we assured to remain in control?

A captivating conversation is taking place about the future of artificial intelligence and what it will/should mean for humanity. There are fascinating controversies where the world's leading experts disagree, such as: AI's future impact on the job market; if/when human-level AI will be developed; whether this will lead to an intelligence explosion; and whether this is something we should welcome or fear. But there are also many examples of of boring pseudo-controversies caused by people misunderstanding and talking past each other. To help ourselves focus on the interesting controversies and open questions — and not on the misunderstandings — let's clear up some of the most common myths.

There have been a number of surveys asking AI researchers how many years from now they think we'll have human-level AI with at least 50%

probability. All these surveys have the same conclusion: the world's leading experts disagree, so we simply don't know. For example, in such a poll of the AI researchers at the 2015 Puerto Rico AI conference, the average (median) answer was by year 2045, but some researchers guessed hundreds of years or more. There's also a related myth that people who worry about AI think it's only a few years away. In fact, most people on record worrying about superhuman AI guess it's still at least decades away. But they argue that as long as we're not 100% sure that it won't happen this century, it's smart to start safety research now to prepare for the eventuality. Many of the safety problems associated with human-level AI are so hard that they may take decades to solve. So, any businessmen ought have business moralty to know whether they ought how to apply their AI to assist their business development in our future office environment to be more moral.

● Five ways to use AI to improve business efficiency to these office tasks

Regardless of a company's size or type, its executives typically look for ways to help it operate as efficiently as possible. They understand the link between efficiency and profitability. If employees waste too much time with drawn-out processes or complicated tasks, it'll be hard for the enterprise to remain profitable and adapt to challenges. Fortunately, artificial intelligence (AI) supports the need for effective business operations. Here are five ways enterprises can use AI for help: 5 ways to use AI to improve business efficiency image.Getting the best results from AI means looking at where bottlenecks exist, then figuring out if and how it might remove or minimise them. AI can help any offices to improve or raise efficiency to these tasks aspects as below:

1. Use AI to answer queries and support customer engagement

Chatbots are an increasingly popular option for businesses to try, and they use AI to work. Companies often build chatbots that can answer any questions from customers that come through outside of business hours. Some identify the nature of a person's problem, then either attempt to tackle it with preprogrammed answers or pass the communications to a human support worker. The retail industry, in particular, saw success by deploying chatbots. Global data collected by Juniper Research shows an estimated 2.6 billion retail-based chatbot interactions in 2019, and the company forecasts the number to rise to 22 billion in 2023.

Chatbots are excellent for answering simple questions like "How late are you open today?" or "Do you have gluten-free menu options?" Getting quick answers to queries like those increases the chances customers will choose to do business with one company over another. Equally importantly, when chatbots can give responses in a matter of seconds, there's no need for humans to stop what they're doing and address the questions.

2. To enhance reporting speed and accuracy

Company reports reveal things such as which products are selling the fastest and where they're most popular. They can also confirm the impacts of marketing campaigns on product sales, break down the costs of a new packaging choice or shipping method, and much more. However, as anyone that files reports knows, creating them is a painstaking task, and trying to rush through the process could cause mistakes. Some forward-thinking companies are combining AI with big data analytics. Doing this brings better forecasts and takes some of the burdens off the people who prepare the reports. AI also helps conquer the inevitability of mistakes. Even the most careful people make blunders, often because of mental fatigue.

AI learns to spot patterns in data and gets smarter with time. This means reports get finished faster and contain more-reliable information. The reliability aspect is crucial, especially since recently published research indicated two-thirds of the senior executives polled had no confidence or trust in big data. Using AI does not mean companies can do without data scientists. However, depending on the technology allows them to reduce the uncertainty that may otherwise exist. It also prevents employees who work with a company's data from being asked to recheck the findings, even if they initially took appropriate precautions to ensure accuracy.

3. To improve data transfer speeds

Fast data transfers help AI technology work. Concerning some information-intensive applications like virtual reality (VR), any slow transmissions greatly interfere with the realism, and content immersion people should enjoy after strapping on a VR headset. As it turns out, AI can improve data transfer speeds, too. For example, services exist that boost speeds across any wide-area network (WAN). Users enjoy consistently accelerated rates regardless of the kind of information transferred. Some companies have solutions that can reduce WAN job times by up to 98%. These AI-driven options work particularly well when companies need to move information between data centres or cloud environments.

4. To assist the IT team with identifying genuine cyberthreats and anomalies

One of the ongoing challenges faced by IT teams of all sizes is to separate the true cyber threats from false alarms. The difficulties associated with categorising the two types may mean cybersecurity professionals waste time getting to the bottom of things that are ultimately nonissues. They might miss the actual threats that could derail a company's operations. Besides detecting possible intrusions associated with a network, AI can screen for software abnormalities that may make it easier for cybercriminals to orchestrate their attacks successfully. It can also find malicious software hackers installed. Due to this kind of information and the advantages of receiving it through real-time updates, IT security teams can work more productively. They can use the majority of their resources on the threats that matter most to the company's stability.

Some organisations have even used AI to help them conquer the substantial skills shortage in the cybersecurity industry. At Texas A&M University, the Security Operations Center deals with about a million attempted hacks each month. The facility has some full-time workers, but students comprise most of the staff. They work alongside AI that aids in threat monitoring, detection and remediation. Before students see possible threats, the smart technology finds and groups them. This approach saves time and lets the team get to work investigating the problems and deciding how to handle them.

5. To streamline the time-to-hire metric when filling new positions

Statistics show the average time required to hire a person for an open position ranges from 12.7 to 49 days, depending on the industry. The timing also varies based on the type of work a job requires. For example, it takes a shorter amount of time overall to find someone for an administrative or human resources position than one associated with a creative or advertising role. Then, of course, interviews are more extensive for high-profile work.

Human resources professionals increasingly use AI to cut down on the time between first posting a job and finding the ideal individual to hire. For example, an AI platform could look for particular desired keywords in submitted resumes, saving hiring managers from poring over the documents themselves. AI can also pitch in during interviews. A company called VCV recently raised $1.7m to further develop its AI tool that has voice and facial recognition components. Candidates are asked to record videos of them answering interview questions, but they can't prepare for the specific content in advance.

In conclusion, AI Can Boost Efficiency at All Types of Companies. The examples here highlight why so many company leaders conclude that if they use AI, they could cut down on inefficiencies. Getting the best results from

AI means looking at where bottlenecks exist, then figuring out if and how it might remove or minimise them. But, AI still lack enough effort to help all staffs to raise efficiency to all department tasks in any office environments.

- How the office energy Department is using AI to solve some of their office staffs electricity toughest challenges in their office working environment.

Insights from artifical intelligence has the potential to transform nearly every aspect of the world as we know it. Today, it is being applied to accelerate the pace of discovery in a wide variety of areas including energy, materials science, health care, national security, emergency response, transportation, and more. AI can be trained to help any energy department to gather data to avoid energy waste to be used to any organizations. So, AI is such as one super machine to do more accurate judgement to help any energy scientists to find the best methods to help any organizations to avoid to waste to use any energy daily. Then, organizations can avoid to spend too much energy to use in offices and they can save more money and avoid energy shortage challenge causes more easily. When the office managers can apply AI ability to reason and put it into a more automated format in a computer system to their every staffs' computer and record their computer electricity use record in their offices every day.

How can AI help offices to save energy or avoid to waste energy ?

The next industrial revolution is already happening. Artificial intelligence (AI) is ushering in an era of technologies that are faster, more adaptable, more efficient, and making the world more digitally connected. AI is best described as complementary to human intelligence, delivering the computing power to crunch numbers too big for people and recognize patterns too tedious for the human eye. In a Harvard Business Review study of 1,500 companies, it was found that the most significant performance improvements were made when humans and machines worked together. As AI becomes one of society's greatest assets, it's especially helpful for solving problems that seem larger than life — like protecting our natural environment.

Through machine learning, robotics, drones, and the internet of things (IoT), society can achieve better monitoring, understanding, and prevention of damage and stressors on Earth's land, air, and water. Even technology already available today could reduce energy usage in the U.S. by 12 to 22 percent, according to The Information Technology Industry Council (ITI). In the face of this dire reality, the potential of technology to help meet this challenge is a rare source of optimism. According to a recent survey by Intel and the research firm Concentrix, 74 percent of business-decision makers working in environmental sustainability agree artificial intelligence (AI) will help solve long-standing environmental challenges; 64 percent agree the Internet of Things (IoT) will help solve these challenges. As the field of AI develops, so will the potential to protect the environment. From the land and air to both drinking and ocean water, AI is shaping up to be the key that governments, organizations, and individuals can tap to work toward a cleaner planet, even AI can help offices to avoid to waste energy when staffs are working in offices every day.

Many AI scientists indicate that AI will also make renewable energy technology like solar panels and wind turbines more efficient and cost effective, helping them to become ubiquitous and lower society's dependence on fossil fuels. AI will also make renewable energy technology like solar panels and wind turbines more efficient and cost effective, helping them to become ubiquitous and lower society's dependence on the fossil fuels polluting the air — then hopefully eliminate them all together. Combined with the smart grid, another technology that will be enabled by AI, this will truly progress the way people receive and use electricity in their homes, offices, and everywhere else. Smart meters save energy by allowing for two-way communication between the grid and anything that uses electricity, giving energy providers a better understanding of usage and the ability to make real-time adjustments for efficiency. Customers will benefit from the real-time data too; seeing the increased costs at peak times will encourage them to voluntarily adjust their usage to save money. This will, in turn, save even more energy: a win-win. Plus, the process of delivering the energy itself will also be improved by the smart grid, thanks to Volt/VAR control systems that can reduce the amount of energy wasted when it's in electricity transmission lines.

Can AI replace office workers

Can AI replace all office workers to do their different tasks in office different department ? If AI can only replace some department office workers to do their simple tasks, how it can raise more efficiency to compare them in some business office environments. I shall indicate some office tasks to explain how AI can help these businesses to raise their efficiency in officesas below:

- AI insurance workers

Nowadays, some country offices begin apply robotics to replace human office workers in their companies. For example, Japanese company replaces office workers with artificial intelligence in insurance industry. A future in which human workers are replaced by machines is about to become a reality at an insurance firm in Japan, where more than 30 employees are being laid off and replaced with an artificial intelligence system that can calculate payouts to policyholders.

Fukoku Mutual Life Insurance believes it will increase productivity by 30% and see a return on its investment in less than two years. The firm said it would save about 140m yen (£1m) a year after the 200m yen (£1.4m) AI system is installed this month. Maintaining it will cost about 15m yen (£100k) a year. The move is unlikely to be welcomed, however, by 34 employees who will be made redundant by the end of March.

The system is based on IBM's Watson Explorer, which, according to the tech firm, possesses "cognitive technology that can think like a human", enabling it to "analyse and interpret all of your data, including unstructured text, images, audio and video".The technology will be able to read tens of thousands of medical certificates and factor in the length of hospital stays, medical histories and any surgical procedures before calculating payouts, according to the Mainichi Shimbun.

While the use of AI will drastically reduce the time needed to calculate Fukoku Mutual's payouts – which reportedly totalled 132,000 during the current financial year – the sums will not be paid until they have been approved by a member of staff, the newspaper said.

Japan's shrinking, ageing population, coupled with its prowess in robot technology, makes it a prime testing ground for AI. According to a 2015 report by the Nomura Research Institute, nearly half of all jobs in Japan could be performed by robots by 2035. For example, one Japan insurance company, Dai-Ichi Life Insurance has already introduced a Watson-based system to assess payments - although it has not cut staff numbers - and Japan Post Insurance is interested in introducing a similar setup, the Mainichi said. AI could soon be playing a role in the country's politics. Next month, the economy, trade and industry ministry will introduce AI on a trial basis to help civil servants draft answers for ministers during cabinet meetings and parliamentary sessions. The ministry hopes AI will help reduce the punishingly long hours bureaucrats spend preparing written answers for ministers.

● AI public service workers

The automated city: do we still need humans to run public services? If the experiment is a success, it could be adopted by other government agencies, according the Jiji news agency. If, for example a question is asked about energy-saving policies, the AI system will provide civil servants with the relevant data and a list of pertinent debating points based on past answers to similar questions.

The march of Japan's AI robots hasn't been entirely glitch-free, however. At the end of last year a team of researchers abandoned an attempt to develop a robot intelligent enough to pass the entrance exam for the prestigious Tokyo University. "AI is not good at answering the type of questions that require an ability to grasp meanings across a broad spectrum," Noriko Arai, a professor at the National Institute of Informatics, told Kyodo news agency. Hence, AI will have possible to replace some public service workers' tasks.

● AI replace warehouse workers

Denso's use of Drishti shows how some jobs will be transformed by artificial intelligence even when they're unlikely to be eliminated by AI anytime soon. Many jobs in manufacturing require dexterity and resourcefulness, for example, in ways that robots and software still can't match. But advances in AI and sensors are providing new ways to digitize manual labor. That gives managers new insights—and potentially leverage—on workers. For example,some workers say the results are unpleasant. Last year, Amazon warehouse employees in Minnesota staged a walkout to protest how the company uses inventory and worker-tracking technology. They allege that Amazon uses it to enforce a punishing working pace that causes injuries. The company has disputed those claims, saying it coaches employees on how to safely meet quotas.

Workers at Denso were initially wary of the prospect of being video-recorded all day to feed machine-learning algorithms, but Huffman says they have since come to appreciate Drishti's technology. After something goes wrong, workers can now look at the data and video with their managers, instead of having to hope bosses take their account of what happened seriously. Huffman says having a constant readout on productivity also helps managers be more responsive to nascent problems. "If somebody's struggling, not every associate is going to call for help," he says. "If we see their cycle time is jumping through the roof, we can go over and say 'Are you having any issues?'"Workers on Denso lines equipped with Drishti's technology now get a personal feed of their own data. Monitors on each workstation display how a worker is doing, says Raja Shembekar, a Denso vice president. If the worker completes their assembly step on time, they see a smiley face—if not, a frowny one. Hence, Amazon had begun to apply AI robotic to replace some warehouse workers' tasks.

For another factory manufacture working environment example, AI can replace many manufacture workers to do their tasks in factories. Route 9 skims by Boston and cuts clear across Massachusetts to Pittsfield, a city of roughly 50,000, the largest in Berkshire County. Well east of Pittsfield, Route 9 becomes Worcester Road, named for a city that in earlier times was the nation's largest manufacturer of wire—barbed wire, electrical wire, telephone wire and the wire used in the making of undergarments by the Royal Worcester Corset Co., once the largest employer of women in the United States. Older Worcester residents can still recall the factory bells pealing to signal the start and end of the workday. Now, the bells are silent, and the wire and corset factories have been replaced with three of the nation's largest employers: Walmart, Target and Home Depot. If this sounds familiar, it should. It has been nearly two decades since retail overtook manufacturing as the nation's most important job creator, employing roughly one of every 10 American workers—more people than in health care and construction combined. That's a lot of jobs.

Of course, not all retail jobs qualify as what most of us consider good jobs. Today, the average hourly wage for a nonsupervisory retail worker is $11.24, and less than half of retail workers receive benefits of any kind. Still, as a nation, we've come to a sort of uneasy peace with this trend. We know that manufacturing employs far fewer Americans today than it once did—that iPads and Macs aren't made in America and neither are many televisions, appliances, tools, toys or clothes. We also know that shopping for these appliances, tools, toys and clothes is an all-American pastime: On average, we spend nearly 45 minutes a day (more than 270 hours per year) purchasing goods and services. Retail has become the world as we know it, and many of us expect to make our living working in that world.Thanks to automation and a killer business model, Amazon is so efficient that it reaps nearly twice the revenue per employee of Walmart, despite the fact that Walmart, too, has a substantial online presence. Worldwide, Amazon has installed over 100,000 robots to labor in "perfect symbiosis" with humans in its warehouses and has plans to install many thousands more. While it's not clear what constitutes perfect symbiosis, the robots are said to save the company $22 million annually, per warehouse. The company's master plan of an autonomous future also includes goods delivered by drones and self-driving vehicles.

For while Amazon continues to open warehouses around the globe and staff them with many thousands of human beings, estimates are that every human on the Amazon payroll—whether full- or part-time—displaces two humans at traditional brick-and-mortar operations. And that's a feature, not a bug: As Tim Lindner, a veteran IT analyst, confided in a note to industry insiders, eradicating jobs is the explicit goal of any online retailer. As he once wrote: "Labor is the highest-cost factor in warehouse operations. It is no secret that Amazon is moving to highly automated operations within its distribution centers, and...it has additional technology that can further reduce the number of humans it needs to process customer orders.... You have heard the old programmer's phrase, 'Garbage in, garbage out.'... [With] the diminishing reading abilities of humans on the Receiving dock, finding an automated solution to eliminate the 'garbage in' problem is the holy grail. Amazon may have just patented it."

By garbage, Lindner meant human error, the alternative to which is apparently robotic precision. And robots can be very precise, especially when it comes to routine tasks. Sawyer, an industrial robot created by the former Boston-based Rethink Robotics, offers an impressive illustration of how all-embracing a robot arm can be. Sawyer is the brainchild of Rodney Brooks, the inventor of both Roomba, the robotic vacuum, and PackBot, the robot used to clear bunkers in Iraq and Afghanistan and at the World Trade Center after 9/11. Unlike Roomba and PackBot, Sawyer looks almost human—it has an animated flat-screen face and wheels where its legs should be. Simply grabbing and adjusting its monkey-like arm and guiding it through a series of motions "teaches" Sawyer whatever repeatable procedure one needs it to get done. The robot can sense and manipulate objects almost as quickly and as fluidly as a human and demands very little in return: While traditional industrial robots require costly engineers and programmers to write and debug their code, a high school dropout can learn to program Sawyer in less than five minutes. Brooks once estimated that, all told, Sawyer (and his older brother, the two-armed Baxter robot) would work for a "wage" equivalent of less than $4 an hour.

Robots loom large in discussions of work and its future, a conversation that can get mired in false assumptions. Until recently, many economists were skeptical that automation could permanently displace human workers on a large scale. People have always shifted away from work better done by machines, but the economic principle of "comparative advantage" predicts that humans will maintain an edge in many fields. Under this logic, technology will not displace us but set us free to do less dangerous, more challenging things, essentially the very things that make humans human. Of course, human workers are complicated. We get tired, hungry, distracted, angry, confused. We make mistakes, sometimes egregious ones. Machines lack our frailties and biases and are better equipped to weigh evidence fairly, without prejudice or false assumptions. Perhaps most critically, machines can retain and process data far more accurately than we can, and that data is growing exponentially.

Every minute of every day, Google services 3.6 million searches in the United States alone. Spammers send 100 million emails. Snapchatters send 527,000 photos, and the Weather Channel broadcasts 18 million forecasts. This and more data—properly collected, codified and analyzed—can be applied to automate almost any high-order task. Data can also serve as a surrogate for human experience and intuition. Online shopping and social media sites "learn" our preferences and use that information to make values-based assessments to influence our decisions and behavior. And, increasingly, machines excel in the tasks once thought uniquely human."Computers are able to see and hear, and have face-recognition capabilities that are significantly better than humans," says Vardi. "Machines understand the human world far better than they did just a few years ago. And we haven't discovered anything in the human brain that can't be modeled."

● AI can replace counter cashier service staffs

And robots need not be perfect, only equal to—or a tad better than—complicated and expensive humans. And technologists are working hard to make sure they are a tad better. For example, in the case of retail, it's become clear that many of us avoid the self-service checkout line—we prefer the cashier to punch in our purchases rather than do so ourselves. So it seems that the job of cashier—among the largest retail employment categories—is not directly at risk. But Zeynep Ton, an MIT management expert who focuses on the retail sector, says self-service checkout is only a first step and not a terribly smart one. "Customers recognized that self-service checkout is not an innovation, but merely a way of outsourcing the job to them, so they didn't like it," she says. "But new technology is coming that will make self-service checkout so much easier and faster, and that will have a real impact on retail employment."

Experts caution that the so-called apocalypse in retail predicted a few years ago has not yet come to pass. In fact, for every company closing existing stores, two more are opening new stores. Retail is a highly competitive industry, and technology is transforming not only the way we shop but the way we connect with brands—for example, just a few years ago, who would have imagined that Amazon would open actual retail stores? And while e-commerce has grown to 10 percent of retail, that still leaves 90 percent for brick-and-mortar stores. But those brick-and-mortar stores, too, are undergoing radical change that has serious implications for America's workforce.

As example, Lobaugh cites food trucks, which he says increasingly pose a threat to many fast-food outlets. Unlike restaurants pinned down by a pair of Golden Arches, food trucks are nimble—they can home in on areas where customers are most likely to gather at any particular time. They can also tailor their offerings to a particular region or even a neighborhood, as well as use Facebook or other media to get out the word on their menu items and locations. Small, specialty stores also have far more flexibility than large department stores. "Technology has reduced the cost of entry into new markets, so in retail there are fewer big, monolithic companies, but more small competitors," he says. "Companies are diversifying to meet the specific needs and desires of consumers—everyone's piece is getting smaller, but there are many more pieces."

But despite what it predicts will be a banner holiday season, this year Amazon took on far fewer seasonal employees than usual—100,000 employees versus 120,000 the previous two years. And while an Amazon spokeswoman insisted that automation is not a factor in this reduced workforce, others seem to not agree. In a recent report, Morgan Stanley analyst Brian Nowak soothed the fears of Amazon shareholders concerned with the wage increase by pointing out that automation had already and would continue to reduce the call for labor, and therefore reduce overall costs. When asked about this, Lobaugh again tactfully declined to comment—other than to say that while the retail sector had lost less ground than most people assume, retail employees were another matter. "There are winners," he says, "and then there are losers."

● AI can replace accountants in accountancy service industry

Not that long ago artificial intelligence (AI), robots and machine learning (ML) were thought to be things only found in science fiction films. Today, this type of technology is taking center stage in workplaces across the globe. Industries, including manufacturing, retail, agriculture, and customer service have already had AI replace some job positions that left workers scrambling to find new career options. This AI revolution is not expected to slow down anytime soon. In fact, experts anticipate that as many as 800 million jobs could be replaced with AI technology by the year 2030. Initially, AI technology and automation in the workplace seemed to only affect pink and blue-collar workers. As this technology advances and becomes more powerful, professional, white-collar workers, including accountants, are starting to worry about what the future holds for their career and if AI will be developed to own their professional skills in accounting service industry.

In basic terms, AI technology is intelligent machines that are able to complete repetitive, mundane tasks at a fraction of the time it takes humans and with greater accuracy. The emergence of Machine Learning now allows AI platforms to observe, analyze and self-learn data and processes to improve its performance and accuracy over time. AI technology is already able to handle many accounting functions, such as tax preparation, payroll, and audits. Many of the leading accounting software providers, including Xero, Intuit and Sage have incorporated AI technology into their software to handle basic accounting tasks, such as bank reconciliations, invoice categorization, risk assessment, and audit processes, like expense submissions and invoice payments. Many of these standard tasks are extremely time-consuming, which has many accountants across the country worried about how the emerging AI technology will affect their billable hours. An even bigger concern is that AI technologies will replace the need for companies to work with accountants at all.

- AI Will Transform not Replace Accountants

While there is no doubt that AI technology is capable of handling many standard accounting tasks faster and more efficiently or that these capabilities will only increase over time, it doesn't mean the end for accountants. There always will be a need for that human element - human intelligence - at the other end of AI technology. In fact, according to leading research firm, Gartner, AI is set to create more jobs than it will replace, leaving workers, including accountants with options. Accountants don't have to worry about their job being replaced by AI any time in the near future. Companies will always need accountants that can analyze and interpret AI data, as well as provide consulting services. Rather than replacing the role of an accountant, AI technology will transform the duties an accountant performs.

With AI technology and machine learning handling many of the mundane, repetitive tasks, accountants will have more time to focus on other aspects of the job, such as consulting and data analysis. This is good news for many accountants. Rather than spending hours completing menial tasks, accountants of the future will be able to use and analyze AI data to provide their clients with sound business solutions.

In many ways, AI will help accountants improve their services. AI technology will improve data entry accuracy and lower the liability risk for accountants. In addition, emerging technology is more efficient at fraud detection, adding an extra layer of protection for accountants and their clients. It also provides real-time data, which allows accountants to provide real-time solutions. Even more impressive is the ability of machine learning to analyze large amounts of data instantly, evaluate past successes and failures in an effort to accurately predict future outcomes.

There is no way to escape the use of AI technology, at least not if you hope to remain competitive in the upcoming years. The speed, efficiency and accuracy of AI technology just cannot be beat. The only thing accountants can do is to embrace this new technology and learn how to maximize its use. The better equipped you are to help your clients integrate and utilize AI technology in their accounting processes the more valuable you will be. For example, many universities today are already incorporating IT and database management courses into their accounting program. This means that graduating students are coming into the workforce with the skills they need for future accounting work. Accountants already in the workforce must find ways to acquire these skills in order to remain relevant to their employers and/or their clients. Accountants can obtain the IT skills they need by attending seminars, using self-learning online programs or attending college-level courses. It is equally important for accountants to stay up-to-date on the latest accounting trends, emerging technologies and industry news. This will allows accountants to not only keep their jobs but to also provide more efficient services to their clients. Rather than worry about AI taking over their jobs, accountants should embrace this technology as a powerful solution to enhance customer services. Finally, accountants will be able to use all their training and experience to provide customer will real and effective business solutions, whether it's in reference to tax consulting, real estate deals, mergers, growth options, or any other business practice.

On conclusion, technology is advancing at record rates so now is the time to obtain the IT and database management skills you need to advance into the future. With the right skills and training, accountants are guaranteed a lucrative career that will last well into the future.

Why Developed And Developing Countries Need Artificial Intelligent Development To Assist Office Tasks

Must developed and developing countries need artificial intelligent development to assist office tasks ? Ought AI is needed to prefer to develop technique to assist office staffs to reduce workload to compare other kinds of occupation environment tasks aspects ? If one developed country, e.g. US, UK , Japan , Singapore it does not continue to develop artificial intelligence, robotic, then what disadvantges or weaknesses , it will encounter to compare when it chooses to continue to develop this artificial intelligent technology in society. If one developing country, e.g. China, Korea, Taiwan, it does not continue to develop artificial intelligence, robotic, then what disadantages or weaknesses, it will also encounter to compare when it chooses to continue to develop this artificial intelligent technology in in society. I shall explan the reasons why the results may cause to either the developed country, or the developing country as below:

- How AI help developing countries to communication and agriculture and learning and medical delivery development

Why can AI help developing countries ? Drones that pick inaccessible crops and mobile phones that give medical advice are two of the ways AI can transform life in the developing world. Artificial intelligence (AI) may improve the lives of the world's poor, the technology needed to revolutionise inefficient, ineffective food and healthcare systems in developing countries is well. For example, in low-income areas, agriculture and healthcare are two critical ecosystems that we can apply AI to immediately; this is not the far future, or even in five years.

Artificial intelligence (AI) has seeped into the daily lives of people in the developed world. From virtual assistants to recommendation engines, AI is in the news, our homes and offices. There is a lot of potential in terms of AI usage, especially in humanitarian areas. The impact could have a multiplier effect in developing countries, where resources are limited.

Emergency Response to developing countries' earthquake natural damage suddence occurrence predicting

AI and machine learning are still finding importance in emerging markets, but certain applications have emerged and are now widely used.

For instance, predictive models for disaster relief enable first responders to automatically analyze large-scale behavior and movement through multiple sources of data including social media platforms, web forums, news sources, etc. Based on collected data, responders can scale reconstruction efforts and distribute supplies in a timely manner.

Why and how AI can assist farmers to predict when the earthquake occurs suddenly in order to avoid or reduce the natural damage to their agriculture productive number loss. For example, In 2015, when a major earthquake hit Nepal, more than 8 million people were affected. During the aftermath, drones were used to map and assess the destruction and speed up the rescue mission. The town of Sankhu, situated about 20 kilometers northeast of Kathmandu, was among the highly affected locations. In May 2018, my company Fusemachines and GeoSpatial Systems partnered with Sankhu's city officials to use drones and artificial intelligence in an effort to automatically estimate the reconstruction need. After processing data accumulated from a drone-powered aerial mapping of the region, the team fed this data to advanced machine learning algorithms. Combining drone imagery, digital mapping and machine learning, the team configured region modeling and infrastructure development with higher accuracy. Another organization known as One Concern, a California-based startup, has created a predictive AI program called Seismic Concern to accurately predict seism and is also working on solutions for wildfires, floods and hurricanes.

Smart AI Agriculture

Another application of AI in developing countries is smart agriculture. Farmers monitor crops more effectively and make better predictions on planting, weeding and harvesting using AI tools. It can also be used to analyze one plant at a time and add pesticides only to infected plants and trees instead of spraying pesticides across large swaths of crops. One California-based tech company is an example of this use of AI. So, the developing countries farmers in rural parts of India are also using AI to increase yields through better access to information about the farming season than they would normally have. Technology-enabled process automation offers the agribusiness industry the chance for remarkable growth -- not only in developed countries but around the world. There's a unique opportunity to increase yields, cut down labor costs and improve people's health.

Medicine Delivery to developing countries' patients urgent need

Companies are also leveraging AI to improve access to health care in some of the most remote areas of the world. In Rwanda, for example, Zipline is using drones to deliver medical supplies and blood to hospitals and clinics that are difficult to access by car. This has dramatically impacted people living in remote parts of the country because they are able to get medical help when needed. The drone system in Rwanda has also helped reduce waste of blood by 95%, as noted by Zipline. One Concern has created an AI program called Seismic Concern that accurately predicts seismic events and is also working on solutions for floods, wildfires and hurricanes. The medical field may actually benefit the most from emerging technologies in developing countries.

Assistance to reduce teaching work workload or psychological pressure to teachers in developing countries‘ schools

Another vital area benefiting from innovative technologies like AI is education. Advanced technologies can enhance how we learn, teach and perform tasks. In most developing countries, schools lack experienced teachers and resources to enhance students' knowledge. As a result, many students still have to walk long distances to get to the nearest school, which has created education gaps, especially in rural areas. AI tools such as personalized learning assistants can simplify learning by making tutoring services and learning materials accessible to all students, wherever they are. Machines can be automated to help students learn basic concepts without a tutor, which companies like Carnegie Learning are working on. This would allow students to learn at any time from anywhere. With AI, education is made easy and accessible to more people.

The initial usage of AI in developing countries has been at a micro level -- solving small, specific problems in a defined industry. As machine learning advances and there is a higher utilization of AI, we will see more complex issues being targeted and resolved. When duly adopted, AI can positively impact future developing countries people everyday lives not just in disaster intervention, education, health care and agriculture but can also help in mitigating poverty, malnutrition and pollution. Especially, in developing nations, to leverage AI's true potential and create a snowball effect. Startups are defining a holistic and humanitarian approach to building more sophisticated, AI-ready societies. Stakeholders in the AI landscape should understand the strengths and nuances of the developing world as well as the limitations of AI and create localized solutions and applications.

Why does smart phone help developing countries communication ?
Internet Seen as Positive Influence on Education but Negative on Morality in Emerging and Developing Nations. Internet access differs substantially across the 32 emerging and developing countries polled, with the lowest rates of internet use in South Asian and sub-Saharan African nations. Within countries, computer owners, young people, the well-educated, the wealthy and those with English language ability are much more likely to access the internet than their counterparts. To access the internet, people increasingly use smartphones rather than more cumbersome fixed landline connections and computers. Around the world, both smartphones and basic-feature phones alike are used for sending messages and taking pictures.

In fact, many developing countries young people, students are popular to use smart phones for internet usage aim, instead of communication. Moreover, many developing countries working people are also popular to use smart phones for any working usage in their working time , even non working time any time. So, smart phones (AI) phones will be important communication or leisure tools to developing countries people in the future. Unless, it is one day, scientists can develop another new communication tool to replace smart phones. So, artificial intelligence will be important to influence developing countries people , how to improve or bring positive learning attitudes to students in their daily learnnng lifes. as well as how to raise developing countries people, how to raise working people efficiency or improve performace in their daily working lifes. So, AI may bring positive learning or working attitudes to developing countries working people and students both.

The Positive Impact of Mass Media in Developing Countries
Radio, newspapers, television, Internet, social media, etc., all of these are forms of mass media. Each of these outlets has the capability of bringing information to thousands of people with one device. While in some communities it is easy to take advantage of these communication outlets such as television and Internet access, not everyone has access to such outlets. Radio is one of the most common forms of mass media in developing countries because it's affordable and uses less electricity than many other forms of mass media, but only approximately 75 percent of people in developing countries have access to a radio, and roughly 77 percent of people in rural areas have access to electricity.

For developing countries that have implemented forms of mass media in their communities, there have been numerous positive outcomes are influenced to impact developing countries mass media by artificial intelligence as below:

When AI is participated to developing countries mass media, it can influence any radio, television audiences raise more attention to each other through social media platforms such as Facebook and Twitter and create, organize and initiate street protests and campaigns. Furthermore, having access to social media in developing countries, people are able to connect to those that they usually wouldn't have the chance to talk to. Moreover, AI Provides educational opportunities- In many countries, the division between local and national languages as well as issues of literacy can make communication difficult. With the use of mass media, a bridge can be built between these two gaps. In India, there is a radio station that provides information in local languages and respects local culture and traditions. One of the main ways is to create public awareness of what is going on with businesses and government officials. The media plays an important role in giving people the opportunity to act against injustice, oppression and misdeeds that they otherwise wouldn't know about. Information on available healthcare, a mass radio broadcast was sent out encouraging parents to seek treatment at local healthcare facilities for their sick children. With this mass outreach on healthcare, the encouragement of people to take their children to healthcare facilities saved thousands of lives. This easy way of encouraging others and bringing awareness about certain diseases was made possible through a simple radio broadcast. Finally, when AI is particiapted to media, it may bring many social issues to life that otherwise would remain unknown to many people. In developing countries and communities like Burkina Faso, when the radio broadcast was released about malaria, diarrhea and pneumonia, people were educated and moved to action and knew to take their children to healthcare facilities for preventative care. As it is seen, having access to different media outlets is vital for those in developing countries. Here are three ways that those in developing countries can implement mass media to help their people and communities.

When AI is participated to any internet radio or internet newspaper mass online listening or reading channel. It can provide online radios or newspapers in public places- By providing online radios and newspapers in public areas it gives community members to access news, information and emergency warnings. Even though radios can be on the cheaper side, there are still many people that can't afford to have a radio in their home. By providing one in a local place, not only would it better educate the community members but also it will bring the community together. So, it can make media outlets a two-way platform- Creating a two-way platform between the community and those who are behind the radio stations, newspapers or broadcasts makes the community feel involved and that their voices are being heard. An organization called Soul City in sub-Saharan Africa is showing how well two-way platforms work by engaging their listeners and having them contribute thoughts and ideas about complex issues. Because developing countries radio listening audiences or newspaper readers are popular to accept computer online radio listening channel or online newspaper reading channel to replace traditional paper newspapers or radio machines. So, AI may raise their listening news or reading news leisure feeling from online mass media channel in the future.

● Why do developed countries need to develop AI

Artificial intelligence, or AI, is driving massive shifts across the globe, and every day more questions arise. What impact will AI have on the workforce and how can we prepare for it? How can we encourage economy-boosting and job-creating technologies? How can we ensure that AI will be implemented ethically and with minimal bias? How will society benefit? For developed country, such as US example. None of the US, Israel and Russia have a formal national AI policy yet. Private sector companies such as Google, Amazon and Apple and the US department of defence are driving the bulk of AI investment in the United States. Though Israel does not have a specific policy, it is keenly focused on AI and has seen the number of AI start-ups triple since 2014.

Developed country may learn whether what weakness it is lacking when it does not continue to develop AI from one another developed country. Which countries are approaching AI most effectively, and to what degree is there opportunity for greater international collaboration? It may be too early to tell; however, when analyzing the best practices of existing national AI policies, there is much that can be learned. These are the specific areas to consider. When one developed country continue to develop or research AI, it may bring these benefits as below:

On gathering Data aspect, from self-driving vehicles to smart cities, data is the driver behind AI. Innovation in the United States is limited without a national strategy that answers questions about protocol and ownership.

France and Denmark, on the other hand, are opening government data. France is hosting troves of centrally collected public and private data that it plans to make available as part of its strategy. Conversely, by taking a restrictive position on issues of data collection (as indicated by the implementation of General Data Protection Regulation), the EU is putting manufacturers and software designers at a disadvantage while balancing the demand for privacy. On raising technologica talent aspect, the demand for AI talent far outweighs the available supply. As a result, almost every nation's strategy addresses talent development. Canada's AI strategy is distinct in that it primarily focuses on research and talent strategy. The country boasts AI degree programmes and is building a $127 million research facility in Toronto. Companies like Facebook and my own company, Uptake, are investing in Canada to access this talent pool. On AI legal technological innovation aspect, a whole host of legal questions swirl around AI. The country is developing a bill for AI liability that will be ready in March 2019. The government hopes the legal framework will attract investors by providing a simple, comprehensive guideline to enable the broad use of AI systems. So, when the developed country applied AI technology to assist any lawyers to work, then AI can help them to reduce the workload to draft any legal documents more easier. So, any developed countries lawyers' draft legal documents time must reduce if the developed countries lawyers accept to apply AI to assist their legal works. One of the great promises of AI is its potential for improving quality of life. But without the right planning and oversight, we risk exacerbating problems of inequality or marginalizing groups of people. As an example, India's AI strategy is focused on leveraging the technology not only for economic growth, but also for social inclusion.

AI may bring what benefits to developed countries

From SIRI to self-driving cars, artificial intelligence (AI) is progressing rapidly. While science fiction often portrays AI as robots with human-like characteristics, AI can encompass anything from Google's search algorithms to IBM's Watson to autonomous weapons. Artificial intelligence today is properly known as narrow AI (or weak AI), in that it is designed to perform a narrow task (e.g. only facial recognition or only internet searches or only driving a car). However, the long-term goal of many researchers is to create general AI (AGI or strong AI). While narrow AI may outperform humans at whatever its specific task is, like playing chess or solving equations, AGI would outperform humans at nearly every cognitive task.

Why research AI safety? Would AI bring war when AI is continued to develop by developed countries? In the near term, the goal of keeping AI's impact on society beneficial motivates research in many areas, from economics and law to technical topics such as verification, validity, security and control. Whereas it may be little more than a minor nuisance if your laptop crashes or gets hacked, it becomes all the more important that an AI system does what you want it to do if it controls your car, your airplane, your pacemaker, your automated trading system or your power grid. Another short-term challenge is preventing a devastating arms race in lethal autonomous weapons.

In the long term, an important question is what will happen if the quest for strong AI succeeds and an AI system becomes better than humans at all cognitive tasks. As pointed out by I.J. Good in 1965, designing smarter AI systems is itself a cognitive task. Such a system could potentially undergo recursive self-improvement, triggering an intelligence explosion leaving human intellect far behind. By inventing revolutionary new technologies, such a superintelligence might help us eradicate war, disease, and poverty, and so the creation of strong AI might be the biggest event in human history. Some experts have expressed concern, though, that it might also be the last, unless we learn to align the goals of the AI with ours before it becomes superintelligent.

There are some who question whether strong AI will ever be achieved, and others who insist that the creation of superintelligent AI is guaranteed to be beneficial. At FLI we recognize both of these possibilities, but also recognize the potential for an artificial intelligence system to intentionally or unintentionally cause great harm. We believe research today will help us better prepare for and prevent such potentially negative consequences in the future, thus enjoying the benefits of AI while avoiding pitfalls.

How can AI be dangerous when developed countries continue to develop AI to become weapon to replace soldiers?

Most researchers agree that a superintelligent AI is unlikely to exhibit human emotions like love or hate, and that there is no reason to expect AI to become intentionally benevolent or malevolent. Instead, when considering how AI might become a risk, experts think two scenarios most likely:

The AI is programmed to do something devastating: Autonomous weapons are artificial intelligence systems that are programmed to kill. In the hands of the wrong person, these weapons could easily cause mass casualties. Moreover, an AI arms race could inadvertently lead to an AI war that also results in mass casualties. To avoid being thwarted by the enemy, these weapons would be designed to be extremely difficult to simply "turn off," so humans could plausibly lose control of such a situation. This risk is one that's present even with narrow AI, but grows as levels of AI intelligence and autonomy increase.

The AI is programmed to do something beneficial, but it develops a destructive method for achieving its goal: This can happen whenever we fail to fully align the AI's goals with ours, which is strikingly difficult. If you ask an obedient intelligent car to take you to the airport as fast as possible, it might get you there chased by helicopters and covered in vomit, doing not what you wanted but literally what you asked for. If a superintelligent system is tasked with a ambitious geoengineering project, it might wreak havoc with our ecosystem as a side effect, and view human attempts to stop it as a threat to be met. So, a super-intelligent AI will be extremely good at accomplishing its goals, and if those goals aren't aligned with ours, we have a problem. You're probably not an evil ant-hater who steps on ants out of malice, but if you're in charge of a hydroelectric green energy project and there's an anthill in the region to be flooded, too bad for the ants. A key goal of AI safety research is to never place humanity in the position of those ants.

Why the recent interest in AI safety ?

Stephen Hawking, Elon Musk, Steve Wozniak, Bill Gates, and many other big names in science and technology have recently expressed concern in the media and via open letters about the risks posed by AI, joined by many leading AI researchers. The idea that the quest for strong AI would ultimately succeed was long thought of as science fiction, centuries or more away. However,, thanks to recent breakthroughs, many AI milestones, which experts viewed as decades away merely five years ago, have now been reached, making many experts take seriously the possibility of superintelligence in our lifetime. While some experts still guess that human-level AI is centuries away, most AI researches at the 2015 Puerto Rico Conference guessed that it would happen before 2060. Since it may take decades to complete the required safety research, it is prudent to start it now.

Because AI has the potential to become more intelligent than any human, we have no surprise way of predicting how it will behave. We can't use past technological developments as much of a basis because we've never created anything that has the ability to, wittingly or unwittingly, outsmart us. The best example of what we could face may be our own evolution. People now control the planet, not because we're the strongest, fastest or biggest, but because we're the smartest. If we're no longer the smartest, are we assured to remain in control?

A captivating conversation is taking place about the future of artificial intelligence and what it will/should mean for humanity. There are fascinating controversies where the world's leading experts disagree, such as: AI's future impact on the job market; if/when human-level AI will be developed; whether this will lead to an intelligence explosion; and whether this is something we should welcome or fear. But there are also many examples of of boring pseudo-controversies caused by people misunderstanding and talking past each other. When one developed country continue to develop AI, can itself country's all factories workers will lose jobs, due to AI can replace them to do simple works in factories, or any public transport drivers, e.g. bus drivers, ferry , tram, train drivers, they will lose jobs, when AI ( non manual driving drivers ) can replace all public transport drivers. So, some occupations will lose if developed countries continue to develop or research AI to replace human to do some simple jobs, such as some cooking jobs can be done by AI. So, it is possible that future cookers won't be needed, because AI cooking skills may be better than them to cook any good taste chinese or western food in restaurants. If you drive down the road, you have a subjective experience of colors, sounds, etc. But does a self-driving car have a subjective experience? Does it feel like anything at all to be a self-driving car? Although this mystery of consciousness is interesting in its own right, it's irrelevant to AI risk. If you get struck by a driverless car, it makes no difference to you whether it subjectively feels conscious. In the same way, what will affect us humans is what superintelligent AI does, not how it subjectively feels.

In fact, AI may be make any brokers jobs in financial market. the main concern of the beneficial-AI movement isn't with robots but with intelligence itself: specifically, intelligence whose goals are misaligned with ours. To cause us trouble, such misaligned superhuman intelligence needs no robotic body, merely an internet connection – this may enable outsmarting financial markets, out-inventing human researchers, out-manipulating human leaders, and developing weapons we cannot even understand. Even if building robots were physically impossible, a super-intelligent and super-wealthy AI could easily pay or manipulate many humans to unwittingly do its bidding. So, future brokers will be replaced by

AI, when AI can be made to own financial brokers‘ analytical mind to make more accurate whether the share price will rise up or fall down to compare human financial brokers' analytical mind. The robot misconception is related to the myth that machines can't control humans. Intelligence enables control: humans control tigers not because we are stronger, but because we are smarter. This means that if we cede our position as smartest on our planet, it's possible that we might also cede control.

Not wasting time on the above-mentioned misconceptions lets us focus on true and interesting controversies where even the experts disagree. What sort of future do you want? Should we develop lethal autonomous weapons? What would you like to happen with job automation? What career advice would you give today's kids? Do you prefer new jobs replacing the old ones, or a jobless society where everyone enjoys a life of leisure and machine-produced wealth? Further down the road, would you like us to create superintelligent life and spread it through our cosmos? Will we control intelligent machines or will they control us? Will intelligent machines replace us, coexist with us, or merge with us? What will it mean to be human in the age of artificial intelligence?

Why do developed countries people need AI ?

Why do we assume that AI will require more and more physical space and more power when human intelligence continuously manages to miniaturize and reduce power consumption of its devices. How low the power needs and how small will the machines be by the time quantum computing becomes reality? Why do we assume that AI will exist as independent machines? If so, and the AI is able to improve its Intelligence by reprogramming itself, will machines driven by slower processors feel threatened, not by mere stupid humans, but by machines with faster processors? What would drive machines to reproduce themselves when there is no biological incentive, pressure or need to do so?

Who says superior AI will need or want to have a physical existence when an immaterial AI could evolve and preserve itself better from external dangers. What will happen if AI developed by competing ideologies, liberalism vs communism, reach maturity at the same time, will they fight for hegemony by trying to destroy each other physically and/or virtually. If AI is programmed to believe in God, and competing AI emerges programmed by muslims, christians or jews, how are the different AI's going to make sense of the different religious beliefs, are we going to have AI religious wars? What if the "powers that be" greatest fear is the emergence of a super AI that police's and rationalizes the distribution of wealth and food. A friendly super AI that is programmed to help humanity by, enforcing the declaration of Human Rights (the US is the only industrialized country that to this day has not signed this declaration) ending corruption and racism and protecting the environment.Most benefits of civilization stem from intelligence, so how can we enhance these benefits with artificial intelligence without being replaced on the job market and perhaps altogether?

Key to the process of machine learning are neural networks. These are brain-inspired networks of interconnected layers of algorithms, called neurons, that feed data into each other, and which can be trained to carry out specific tasks by modifying the importance attributed to input data as it passes between the layers. During training of these neural networks, the weights attached to different inputs will continue to be varied until the output from the neural network is very close to what is desired, at which point the network will have 'learned' how to carry out a particular task. A subset of machine learning is deep learning, where neural networks are expanded into sprawling networks with a huge number of layers that are trained using massive amounts of data. It is these deep neural networks that have fuelled the current leap forward in the ability of computers to carry out task like speech recognition and computer vision.

In conclusion, when developed countries continue to develop AI, it may bring positive advantages to bring raising productivies, or efficiencies, but it may also raise unemployment ratio to any low skill or low knowledge jobs in ther societies. However, human future society will need to change to be better to raise our living standard. But AI is one kind the best choice tool to achieve this aim in our future, so I agree developed countries continue to develop or research AI to be the super -human machine.

Artificial Intelligence Worker Brings

Working Environment Influences

Robots were once known only for the manufacturing business but today they are very much part of many workplaces. The future is even more promising for this wonder of artificial intelligence.Imagine a robot doing some of the major tasks of managers like using data to evaluate problems, making better decisions, monitoring team performance, and even setting goals.

Technology is playing a pivotal role in helping humans work more effectively. Since automation has become an integral part of business operations, we can predict that robots are soon going to replace many jobs that are today performed by humans. Now that the corporate world is also on the cusp of entering the robotic age, let's see what pros and cons this technology offers business world. If one day, our global working environments have any kinds of robotic participates to our service and warehouse and office etc. different working environment in order to assist office workers, service workers, warehouse workers, professional lawyers, doctors accountants job duties, what positive or negative influences, it will bring to what negative or positive effects to any office , warehouse, shopping center, hospital, transport , restaurant etc, different working environments. Can robotic help office , warehouse to raise efficiency ? Can robotic help hospital, restaurant, cinema, shopping center to improve service performance? Can robotic influence working environment to be worse? Can robotic help office or any working places to reduce expenditure or reduce long time machine and salary cost when they do not need more employees or machines , due to robotic workers assistance.

I shall attempt to explain whether robotic workers will bring what positive or negative influence to our future working environment as below:

Advantages to robotic bring to working environment

What advantages that robotic will bring to working environment? They may include: Many people fear that robots or full automation may someday take their jobs, but this is simply not the case. Robots bring more advantages than disadvantages to the workplace. They enrich a company's ability to succeed while improving the lives of real, human employees who are still needed to keep operations running smoothly. If you're thinking about investing in some robots, share the advantages with your employees. You might be surprised at how many of them are quick to support the idea.

1. Safety

Safety is the most obvious advantage of utilizing robotics. Heavy machinery, machinery that runs at hot temperature, and sharp objects can easily injure a human being. By delegating dangerous tasks to a robot, you're more likely to look at a repair bill than a serious medical bill or a lawsuit. Employees who work dangerous jobs will be thankful that robots can remove some of the risks.

2. Speed

Robots don't get distracted or need to take breaks. They don't request vacation time or ask to leave an hour early. A robot will never feel stressed out and start running slower. They also don't need to be invited to employee meetings or training session. Robots can work all the time, and this speeds up production. They keep your employees from having to overwork themselves to meet high pressure deadlines or seemingly impossible standards.

3. Consistency

Robots never need to divide their attention between a multitude of things. Their work is never contingent on the work of other people. They won't have unexpected emergencies, and they won't need to be relocated to complete a different time sensitive task. They're always there, and they're doing what they're supposed to do. Automation is typically far more reliable than human labor.

4. Perfection

Robots will always deliver quality. Since they're programmed for precise, repetitive motion, they're less likely to make mistakes. In some ways, robots are simultaneously an employee and a quality control system. A lack of quirks and preferences, combined with the eliminated possibility of human error, will create a predictably perfect product every time.

5. Happier Employees

Since robots are often assigned to perform tasks that people don't particularly enjoy, like menial work, repetitive motion, or dangerous jobs, your employees are more likely to be happy. They'll be focusing on more engaging work that's less likely to grind down their nerves. They might want to take advantage of additional educational opportunities, utilize your employee wellness program, or participate in an innovative workplace project. They'll be happy to let the robots do the work that leaves them feeling burned out.

6. Job Creation

Robots don't take jobs away. They merely change the jobs that exist. Robots need people for monitoring and supervision. The more robots we need, the more people we'll need to build those robots. By training your employees to work with robots, you're giving them a reason to stay motivated in their position with your company. They'll be there for the advancements and they'll have the unique opportunity to develop a new set of tech or engineering related skills.

7. Productivity

Robots can't do everything. Some jobs absolutely need to be completed by a human. If your human employees aren't caught up doing the things that could have easily be left for robots, they'll be available and productive. They can talk to customers, answer emails and social media comments, help with branding and marketing, and sell products. You'll be amazed at how much they can accomplish when the grunt work isn't weighing them down.

8. Cost reduce

The first and the foremost advantage of having robots in workplaces is their cost. Robots are much cheaper than humans and their cost is now decreasing. It's a fact that we cannot compare human abilities with robots but robotic capabilities are now growing quickly. For example, if you run an essay writing service, you can use robots to perform every kind of research related to any subject. Because robots are more active and don't get tired like humans, the collaboration between humans and robots is reducing absenteeism. The pace of human cannot increase hence robots are helping humans.

However,robots are more precise than humans; they don't tremble or shake as human hands. Robots have smaller and versatile moving parts which help them in performing tasks with more accuracy than humans. There is no doubt that robots are significantly stronger and faster than humans. Robots come in any shape and size, depending upon the need of the task. Robots can work anywhere in any environmental condition whether it is space, underwater, in extreme heat or wind etc. Robots can be used everywhere where human safety is a huge concern. Robots are programmed by a human; they cannot say no to anything and can be used for any dangerous and unwanted work where humans may deny to offer their services. For example, many robotic probes have been sent into space but have never returned. Robots in warfare are saving more lives and have now proven to be very successful. For example, in chemical factory environment, robots are now being used in the chemical industry and can, for example deal with chemical spills in a nuclear plant, which would otherwise pose a major health concern. Cost-effectiveness is one of the most sound arguments to be made for the case of industrial robots. Robots will reduce production costs by eliminating internal costs to compensate human salaries. Businesses are forecasting that their profitability will increase once they implement robots into production, or that they will have more financial mobility to invest in new products or technologies.

9. productive efficiency

Quality assurance is expected with the use of machinery in production. Industrial robots will be able to ensure consistency with mass production of manufactured products. The possible human error that assembly line workers pose the threat of will be removed. Optimized production efficiency means that a general manager will be able to have set quantity and quality standards that will be met by robots. Production quotas will not be jeopardized by low concentration, break time and employee injuries, among other things. The efficiency of production forecasts and supply levels will be increased with robots, able to be programmed to work at the optimal speed for a given plant. Limiting human work in hazardous environments, because manufacturing jobs often place workers at more physical risk compared to a lot of other industries. Lowering the level of a hazard presented to employees on the job is attractive to executives to preserve company reputation and minimize potential legal liabilities.

10. Reducing longer working hours

Typically people have to have breaks, get distracted and after time attention drops and pace slows. With a robot it can work 24/7, and keeps running at 100%. Typically if you replace one person on a key process in a production line with a robot the output increases by 40% in the same working hours just because a robot has more stamina and never stops. Robots also don't take holidays or have unexpected days off sick.

11. Increased profitability

By increasing the efficiency of your production process, reducing the resource and time needed to complete it, and also achieving higher quality products, industrial robots can thus be used to achieve higher profitability levels overall, with lower cost per product.

12. Improved working environment

Industrial robots are often used for performing tasks which are deemed as dangerous for humans, as well as being able to perform highly laborious and repetitive tasks. Overall, by using industrial robots you can improve the working conditions and safety in your factory or production process. Robots don't get tired and make dangerous mistakes, neither do they suffer from repetitive strain injury.Due to their high accuracy levels, robots can also be used to produce higher quality products which adhere to certain standards of quality, whilst also reducing the time needed for quality control.Industrial robots are able to complete certain tasks faster and better than people, as they are designed to perform these tasks with a higher accuracy level. This and the fact that they are used to automate processes which previously might have taken significantly more time and resources, means that you can often use industrial robots to increase the efficiency of your production line.

13. Improved Quality Assurance

Few workers enjoy doing repetitive tasks and after a certain period of time concentration levels will naturally decline. This lapse in concentration is known as vigilance decrement and can often lead to costly errors for the business and sometimes serious injury to the member of staff.Robotic automation eliminates these risks by accurately producing and checking items meet the required standard without fail. With more product going out the door manufactured to a higher standard, this creates a number of new business possibilities for companies to expand upon.

14. Increased Productivity

Using robotic automation to tackle repetitive tasks makes complete sense. Robots are designed to make repetitive movements Humans, also by design, are not. The introduction of automation into your manufacturing process has many different productivity benefits, some of which are shown here.Giving staff members the opportunity to expand on their skills and work in other areas will create a better environment which the business as a whole will benefit from. With higher energy levels and more focus put into their work, the product can only improve, which will also lead to extremely satisfied clients.

15. Avoiding workers need to work In Hazardous Environments

Aside from potential injuries in the workplace, staff members in particular industries can be asked to work in unstable or dangerous environments. For example, if a high level of chemicals are present, robotic automation offers the ideal solution, as it will continue to work without harm. Production areas that require extremely high or low temperatures typically have a high turnover of staff due to the nature of the work. Automated robots can minimise material waste and remove the need for humans to put themselves at unnecessary risk.

Disadvantages to robotic bring to working environment

1. Increase unemployment rate and job loss

On working environment cost increasing aspect, where robots are increasing the efficiency in many businesses, they are also increasing the unemployment rate. Because of robots, human labour is no longer required in many factories and manufacturing plants. They can certainly handle their prescribed tasks, but they typically cannot handle unexpected situations.The ROI of your business may suffer if your operation relies on too many robots. They have higher expenses than humans, so at the end of the day you may not always achieve the desired ROI.

However, robots may have AI but they are certainly not as intelligent as humans. They can never improve their jobs outside the pre-defined programming because they simply cannot think for themselves. Robots installed in workplaces still require manual labour attached to them. Training those employees on how to work with the robots definitely has a cost attached to it.Moreover, robots have no sense of emotions or conscience. They lack empathy and this is one major disadvantage of having an emotionless workplace.Also, robots operate on the basis of information fed to them through a chip. If one thing goes wrong the entire company bears the loss. Where a robot saves times, on the other hand it can also result in a lag. It is, after all, a machine so you cannot expect too much from them. If a robot malfunctions, you need extra time to fix it, which would require reprogramming.If ultimately robots would do all the work, and the humans will just sit and monitor them, health hazards will increase rapidly. Obesity will be on top of the list. So there are advantages, but there are disadvantages as well. It is the twenty first century and we cannot work without machines.Humans are still considered far more efficient than robots when it comes to decision making powers, handling difficult situations, brainstorming, and generally bringing a sense of emotion and empathy into a workplace. Besides, you cannot rule out the significant role of humans in a business. After all, no machine can replace the human factor 'real employees' bring into a workplace. So, AI can raise unemployment and increase factory or shopping center or office working environment cost when their working environment are applied robotic to replace many workers, then machine electricity expense will also increase. Otherwise, human workers can not spend too much electricity expense in cost aspect.

Whilst industrial robots can prove highly effective and bring you a positive ROI, implementing them might require a fairly high capital cost. That's why, before making a decision we recommend considering both the investment needed and also the ROI you expect to achieve. Often the easiest way to get round this issue is to take out asset finance and the ROI of the robot more than pays for the interest on the asset finance.

This is typically the biggest obstacle that will decide whether or not a company will invest in robotic automation, or wait until a later stage. A comprehensive business case must be built when considering the implementation of this technology. The returns can be substantial and quite often occur within a short space of time. However, the cash flow must be sustainable in the meantime and the stability of the company is by no means worth the risk if the returns are only marginal. Yet, in most instances there will be a repayment schedule available, which makes it a lot easier to afford and control finances. Our downloadable automation payback calculator also has a finance scheme option so you can see how this would work for you.

On job loss increasing aspect, Job loss is by far the most significant opposition frequently brought against the use of robots in the manufacturing industry. Industry workers of all levels, from entry-level to veterans, worry about the security of their employment status, and the ability of their job to be replaced by a robot. This panic is more widespread in this industry compared to others because of the closer immanence of a robot takeover in manufacturing.

Macro effects are another topic that usually comes up with job loss. More “big picture” thinkers wonder how the national, and eventually global economy will be affected when manufacturing workers’ jobs are displaced. How can this mass unemployment possibly be compensated for, and how can the robots’ presumed success be limited from seeping into other industries. However, increased investment costs are a financial counterpoint to industrial robots, with the idea that manufacturing companies will rack up their debt investing in robotic technology. Firms that do not have the funding might even go bankrupt in an effort to keep up with industry trends rather than continue on with normalized operations.Hence, elimination of a whole labor class would presumably occur a bit of a ways down the road, but the implications of this point are too large not to consider. Bringing in robots to take unskilled labor jobs will place more pressure on the economy, education system, and financial market, just to name a few. The United States has always been associated with the grit and work ethic of its blue-collar workers, and robots are threatening to eliminate this aspect of the human population, with a take over of production jobs.

One of the biggest concerns surrounding the introduction of robotic automation is the impact of jobs for workers. If a robot can perform at a faster, more consistent rate, then the fear is that humans may not be needed at all. While these worries are understandable, they are not really accurate.The same was said during the early years of the industrial revolution, and as history has showed us, humans continued to play an essential role. Amazon are a great example of this. The employment rate has grown rapidly during a period where they have gone from using around

1,000 robots to over 45,000

2. Robotic can not perform better to compare human workers, when they need to work long time in any working environment

Robots need a supply of power, The people can lose jobs in factories, They need maintenance to keep them running, It costs a lot of money to make or buy robots, The software and the equipment that you need to use with the robot cost much money. Robots cost much money in maintenance & repair, The programs need to be updated to suit the changing requirements, the machines need to be made smarter, In case of breakdown, the cost of repair may be very high, The procedures to restore lost code or data may be time-consuming & costly.

Robots can store large amounts of data but the storage, access, retrieval is not as effective as the human brain, They can perform repetitive tasks for a long time but they do not get better with experience such as the humans do. Robots are not able to act any different from what they are programmed to do, With the heavy application of robots, the humans may become overly dependent on the machines, losing their mental capacities, If the control of robots goes in the wrong hands, Robots may cause the destruction. Robots are not intelligent or sentient, They can never improve the results of their jobs outside of their predefined programming, They do not think, They do not have emotions or conscience, This limits how the robots can help & interact with people. Robots can take the place of many humans in factories, So, the people have to find new jobs or be retrained, They can take the place of the humans in several situations, If the robots begin to replace the humans in every field, They will lead to unemployment.

Humans fear robots, Robots inspire two types of fear: firstly, that they might take over our jobs, and secondly, that they could take over the world, Robots will steal our jobs, Robots have the effect of increasing productivity rather than eliminating jobs.Robotics become increasingly present in our everyday life, with household robots, medical, industrial, on production lines, not to mention airports, banks, and hotels, So, Robots may dominate the human species. Robots can operate on the basis of information fed to them through a chip, when one thing goes wrong the entire company bears a loss.The robot can save times, but it can also result in a lag, It is a machine so you can't expect too much from them, If the robot has malfunctioned, you need extra time to fix it, which would require reprogramming, If robots would do all the work, and the humans will just sit and monitor them, health hazards will increase rapidly, Obesity will be on top of the list and less labour at workplaces.

3. Increasing training expense

Whilst industrial robots are excellent for performing many tasks, as with any other type of technology, they require more training and expertise to initially set up. The expertise of a good automation company with a support package will be very important. To minimise your reliance on automation companies you can train some of your engineers present in our everyday life, with household robots, medical, industrial, on production lines, not to mention airports, banks, and hotels, So, Robots may dominate the human species. Robots can operate on the basis of information fed to them through a chip, when one thing goes wrong the entire company bears a loss.The robot can save times, but it can also result in a lag, It is a machine so you can't expect too much from them, If the robot has malfunctioned, you need extra time to fix it, which would require reprogramming, If robots would do all the work, and the humans will just sit and monitor them, health hazards will increase rapidly, Obesity will be on top of the list and less labour at workplaces.

3. Increasing training expense

Whilst industrial robots are excellent for performing many tasks, as with any other type of technology, they require more training and expertise to initially set up. The expertise of a good automation company with a support package will be very important. To minimise your reliance on automation companies you can train some of your engineers

on how to program robots, but you will still need the assistance of experienced automation companies for the original integration of the robot. In recent years the number of industrial robots and the applications they can be used for has increased significantly. However, there still are some limitations in terms of the type of tasks they can perform, which is why we suggest that an automation company looks at your requirement to assess the options first. Sometimes a bespoke automated system may give a better or faster result than a robot. Also, a robot does not have everything built into it, often the success or failure of an industrial robotic system depends on how well the surrounding systems are integrated e.g. grippers, vision systems, conveyor systems etc. Only use good trusted robot integrators to be sure of the optimum results if you do choose to use industrial robots.

Artificial intelligent working participation how brings economic growth

● How AI impacts economy

What is AI non-manual shops? Why and how it can assist economic growth? I shall explain as below:

Artificial Intelligence, as we see it, is a collection of multiple technologies that enable machines to sense, comprehend and act—and learn, either on their own or to augment human activities. Compelling data reveal a discouraging truth about growth today. There has been a marked decline in the ability of traditional levers of production—capital investment and labor—to propel economic growth. Artificial intelligence (AI) is a new factor of production and has the potential to introduce new sources of growth, changing how work is done and reinforcing the role of people to drive growth in business. Accenture research on the impact of AI in 12 developed economies reveals that AI could double annual economic growth rates in 2035 by changing the nature of work and creating a new relationship between man and machine. The impact of AI technologies on business is projected to increase labor productivity by up to 40 percent and enable people to make more efficient use of their time.

To fulfill the promise of AI as a new factor of production that can reignite growth, Accenture recommends the following steps be taken to help navigate the complexity of issues:

Prepare the next generation – integrate human intelligence with machine intelligence so they can successfully co-exist in a two-way learning relationship and reevaluate the type of knowledge and skills required for the future.

Encourage AI-powered regulation – update and create adaptive, self-improving laws to close the gap between the pace of technological change and the pace of regulatory response.

Advocate a code of ethics for AI – ethical debates should be supplemented by tangible standards and best practices in the development and use of intelligent machines. Address the redistribution effects – policymakers should highlight how AI can result in tangible benefits and preemptively address any perceived downsides of AI, helping groups disproportionately affected by changes of employment and incomes.

Our research strongly shows that AI can unleash remarkable benefits across countries, countering slow economic growth and lagging productivity. To fulfill the promise of AI, relevant stakeholders must be thoroughly prepared – intellectually, technologically, politically, ethically and socially - to address the benefits and challenges that can arise as artificial intelligence becomes more integrated in our daily lives."

How AI impacts economy? AI has the potential to markedly increase industry growth. Information and Communication, Manufacturing and Financial Services are the three sectors that will benefit most from the application of AI. AI offers unprecedented profitability opportunities. For example, manufacturing has a forecast share-of-profit increase of 39 percent due to AI-powered systems whose ability to learn, adapt and evolve over time can eliminate faulty machines and idle equipment.

CHAPTER SIX

# Technology or human behavior whether may influence economic growth or recession

Human Behavioral network job brings social economic benefits

What does human network job mean ? Why may human network job be popular? Why human network job behavior may influence economy ? Nowadays internet is popular to use. We can apply internet to find data , search any new things, even earn money. Why does internet may become huma network job source. For example, e-publish may be one kind of new human network job. Any authors may apply internet channel to help them to sell electronic or paper books from e-publisher web store. They may apply facebook, you tub etc. any online channel to promote themselves new books to let new readers to know whether when they may buy themselves favourable new topic books to read from electronic publisher web store.

Thus, future electronic publisher industry may help any authors to build internet network platform to help them to sell and promote ot advertise their any one new electronic or paper book topic to let global any one reader to choose to buy their any new topic books from electronic publisher web store easily and conveniently. However, it implies that electronic network platform author may be one kind of future new human network job in our societies.

How electronic network platform author job may bring economy benefit in macro economy view? A person can have few friends, contacts and still be very influential if these few

friends and contacts are themselves highly influential, e.g. one author must not need to know any one reader in global society. When they like to choose any electronic books from electronic internet network platform. They may become the author's any one topic book buyer, when they feel the author's any one topic book is fun and attract they make decision to buth the strange author whose the topic book from electronic book publisher's platform web store conventiently in short time. Although, they are strangers, they do not know themselves , but the reader can understand what it way that made Google from writing platofrm to create new creative mind and typing network job method to replace traditional hand writing book method for global authors. It will be one kind of new human network writing job.

Hence, global any one reader can apply an innovative search engine , such as google.com to find whether whom author personal new topic books are value to read from internet.

Then, the electroniuc publisher's web store may be new book store platform sale network to help the author to sell many electronic or paper books from electronic network platform

in short time. So, internet may be future new network plaform to help global any one author to create network writing job absolutely. Furthermore, internet may be popular social media

to help any one author to build goold relationship between his/her readers. It is one kind of new network, human network job. New authors do not need to buy many paper books to prepare to put in any one book shop warehouse. Their every book can print on demand to reduce out of book stock in any one book shop. They may choose to sell either electronic books or paper books both from any one book publisher web store. So, electronic network platform may be one kind of good writing channel to help human authors to create income and it can also help authors to bring new creative mind and new topic fun content books to let readers to know and buy to read from electronic publisher network platform.

Why does human behavior may be one kind of new human network job to bring global economic advantages. ALthough, it may be free income or without inocme, but the person does the network behavior, his/her behavior may be bring advantages to influence many other people's health. For this case, when a worker in a coffee shop in an airport gets a vaccination aganinst the flu, it does not only helps him or her stay healthy, but also helps the many travellers who might otherwise have been inflected if that workers caught the flu. So, the externality , the result implies the vaccination of even a part of a community conveys benefits to the whole community. For example, governments pay special attention to the vaccinations of school children, teachers, health mothers, and the elderly, categories of people particularly susceptible not only to catching, but also to transmitting a disease.

It is not accidential that governments are heavily involved with vaccination . When there are externalities, free market, fail to persuade individual incentives with society's

their the worker's decision of whether to get a vaccine ends up attracting whether other people get sick. The workers might not fully take all these other people's potential suffering into account when making her or his vaccination decision.

As Stanford University does many suggestions, understand this and tries to help them make the right decisions and so providers free flu vaccines for its staff and students.

Small pockets of unvaccinated individuals can allow a disease to gain a spread more widely well-being. For example, parent weighing the costs and benefits of a vaccine for their child is not always thinking of the consequences of that vaccination to other people. THese are markets in which subsidizing or regulating behavior can make everyone better off. Because the reason for requiring that a child be vaccinated before enrolling in school is not just to protect that child, because each child's vaccination affects others via potential contagions.

Robots take our jobs behavioral and economy influences

Robot job behavior brings economy influences

If one day robots can replace human to do simple, even complex jobs. They will bring what influences to our global societial economy.The popular economic refrain declares that the

global middle class is dying and robots will soon take our jobs, e.g. shopping center customer service jobs, library service jobs, cinema ticket sale jobs, restaurant kitchen cooker jobs,

even, bus drivers, taxi drivers etc. public transport driving jobs, accountant, doctors etc. professional jobs. Whether it is beautiful or petty matter if our future societies have many human jobs can be replaced to do from robots.

Businessman must may reduce to employ employees and reduce to pay salary or wage, when robots can be replaced to do their employees tasks. But, societies must bring unemployement rate rises , due to societies will have many people loss jobs when their employers choose to buy robots to serve their clients or do any office tasks or customer service or cleaning etc. tasks.

In micro economy view, employers may save money in long term, but in macro economy view, it will cause unemployment ratio rises , even crime rate rises when there are many people lose
jobs in societies. These models of doom, though, fail to account for the hundreds of businesses riding the waves of change in their industries when robots may be invented to replace human to do many simple , even complex tasks in our future societies.

WE may image that one small factory needs to manufacture fishes canes to sell to supermarket, the small , cheaper stuff and higher margin parts of the fishes manufacture industry. Before, this factory needs to employe many human factory workers need to help every fresh customer makeing the perfect fishing gear, designed for performance, durability, and cost in order to achieve to manufacture every fish cane in whole fished processing manufacturing stages. Every worker needs to spend about 15 to twenty minutes to finish every fish cane , till to delivery to any supermarket to sell. If this fish canes manufacturing factory can apply manufacturing robots to help them to finish any one working tasks , every robot can only spend five minutes to finish whole fresh fish cane manufacturing process. Thus, every robot can
help this factory save 10 to 15 minutes time to finsh every fish cane manufacturing process. IN fact, time is money, because when every robot can help this factory to reduce 10 to 15 minutes time to compare human worker. Then, this factory can finish about 20 fish canes in one hour if it can use robot to help it to manufacture fish canes. Otherwise, if this factory still use human workers to help it to manufacture fish canes, then it can finsh about 3 to 4 fish canes in one hour. SO, the manufacturing efficiency ensures that robots must help this fish manufacturing factory to raise fish canes number more than human workers. So, in robotic behavioral economy view, manufacturing robots must help this fish canes manufacturing factory to raise fish canes manufacturing number and deliver increasing number to supermarkets to prepare to sell every day. Robots can help this fish canes manufacturing factory bring manufacturing time saving,

rising manufacturing efficiency, improving performance and reducing wages expenditure long time advantages in micro economy view. However, manufacturing robots can also bring disadvanages to society, e.g. increasing unemployment ratio, increasing crime rate,
this factory workers will lose jobs and income, they need earn social welfare from government and increasing government finance pressure in short time, even long time in macro economic view.

Stanford University graduate program in economics, Scott lecturer explained that "in demand and supply economic theory for robots supply and demand case, robots supply number increasing may influence human workers demand number decrease. It sometimes calls " the efficient frontier".
No specific human beings were mentioned in any of economics classes. As robots supply and demand in market case, They ( robots) may be purely theoretical " agents" who reached to the most reasonable sale prices in order to persuade any one businessman buyer to make manufacturing robot buying decision whether robots can help him / her to bring how much saving time , saving money, saving cost, improving performance, efficiency economic benefit before he/she plans to reduce workers number when he/ she decides to apply robots to replace human workers in his/her factory or office or any service department, e.g. cinema ticket sale service, shopping center customer service, shopping center cleaning , supermarket customer service etc. service or sale tasks. When robots can replace human to do any one of these tasks in any organizations. So, robots may be human worker agents who reached to prices the way robots would react to a software
command. There was nothing that explained why some people thrived and others did n't or why truly brilliant, hardworking people could fail when much lazier folks succeeded." Having been admitted to the Stanford University graduate program in economics, Scott lecturer hoped to get his answers there.

How robots influence our future social changing? Using the right technology can be a boon to your business in this economy. For internet example, it is easier than ever to find well-matched customers all around the world, to stay in contact with them, and to more quickly design the products they want. If you focus solely on being cutting -edge, though you risk letting the technology
take over what should be very robust relationships with your customers , employees, and colleagues. IN nowaddays society, technoligical advances

and cutomation, personal
relationships in business are more crucial than ever. I mean that robots can not replace human to serve clients to let them to feel more comfortable and passion more easily. For shoe shop case example, if the shoe shop apply one robot to serve its clients to replace human shoe salesperson to serve its shoe customers. Robots ensure that they can not persuade every shoe potential buyer to make shoe buying decision more easily when robots need to contact every shoe potential buyer. The reason is simple, because robots can not touch any one shoe buyer individual emotion very easier.
If the shoe buyer needs the robots to help him/her to choose any right shoe styles when he/she can not feel himself / herself can make the most right shoe style choice decision. The robots can not replace human shoe salesperson to make shoe style choice judgement more easily. They must need longer time to analyze whether which shoe style may be the most suitable to the shoe buyer. Otherwise, human shoe salesperson may attempt to make the most right shoe style choice decision to help any one shoe buyer to chooce the most right style shoe because he/she owns shoe style sale experience, shoe style knowledge, the most important reason is that they can feel every shoe customer individual emotion to touch whether he/she will feel comfortable or happy when they attempt to help every shoe customer to seek the most right shoe style in every shoe customer whole shoe searching processing. Othwerwise, serving robots are only one machine, they can not touch or feel every shoe customer individual emotion whether he/she feel comfortable or unhappy or happy when they need to contact them in whole shoe searching processing. Hence, I believe that some tasks robots can
not repalce human staff to do very easily. Otherwise, robots may bring disadvanatges to let any one businessman to loss his/her customers, due to robots can not touch every customer
emotion to compare human staff in service tasks more easily. Robots serving customer behaviors may cause money lose and customers number lose to the shop in micro economic view.

Intellectual human economic behaviors

What does intellectual human economic behaviors mean ? I believe that when we choose or decide to do intellectual behaviors, then our societies will be influenced to bring economic growth in consequence.I shall attempt to indicate pollution case to explain how and why eithet our intellectual or foolish behaviors may bring economic growth or recession in consequence as below:

On one hand, for air pollution social case aspect example, if we only consider to buy cars to drive for working aimr or holiday leisure aim. Then, our societies air will be polluted. Our health will be influenced to bad. Our car driving behaviors may cause global environment air pollution serously. In long tiem, global air pollution will bring our bodies health to be bad. Although, ourselves car driving behaviors may bring our driving travelling leisure enjoyment and comfortable feeling in short time, also we so not need to pay public transport fare often, but we need to compensate ourselves health economic intangible loss due to air pollution , when cars number increases, dirty air will cause ouselves health to become bad.

In the result, we will need to pay more medical expenditure when we are old age, due to ourselves bodies will become bad, due to we breathe global dirty air every day, due to ourselves cars pollute air in long time, e.g. 10 to 20 years, even 30 more without limited air pollution environment. So, driving cars behavior may be one kind of human foolish behavior and our foolish behavior may bring ourselves future long time medical expenditure absolutely.

One the other hand, water pollution social aspect, if we often keep much rubblish to pollute sea, oil exploration porcessing pollute ocean , ships gas pollute ocaen, then fishes will eat polluted food and drive dirty water, due to global ocean is polluted.

In fact, because human only to conside how to buy boats to carry on leisure enjoyment activities, or catch cruises to travel on the sea. Also, oil manufacturers only consider researching anywhere to find new oil exploration places to manufacture oil product, when their oil exploration processes pollute ocarn . Consequently, global fishes drink polluted warer or eat polluted food. They will have poison. SO, human will have high chance to eat poison polluted fishes, due to fishes are poison or are polluted. So, human is doing foolish activities, we only hope to find oil exploration places to pollute ocean or we only spend money to buy ticket to catch ships to travel anywhere in global ocean. All of these human foolish behaviors will bring pollution to global ocean. On consequently, we will need to compensate to eat polluted or dirty or poision fishes, ourselves bodies health will be bad. In long time, we need have high chance to pay medical expenditure when we are old. So, pollution case may be one good example to explain how and why human foolish behavior may influence ourselves future need to compensate serious medical loss.

All of these human foolish behavior will bring pollution to global ocean. On consequently, we will need to compensate to eat polluted or dirty or poison fished , ourselves bodies health will be bad. In long time, we will have high chance to pay medical expenditure, when we are old. So, pollution case may be one good example to explain how and why human ourselves intellectual or foolish behaviors may influence future long time economic loss or economic growth or recession in micro and micro economic view.

On another water pollution aspect hand, if we often keep rubbish to sea, oil exploration processing pollutes ocean and ships' gas pollute ocean, then fishes will eat polluted food and drink dirty water, due to fishes will eat polluted food and drink dirty sea water because the global ocean is polluted seriously.

In fact, because human only consider how to buy boats to carry on any leisure water activities, or catches cruises to travel on the sea. Also, oil manufacturers only consider any where to find oil exploratin places to manufacture oil products from ocean, when their pol exploration processes can plooute ocean. Consequently, global fishes drink polluted water or eat direty food. They will have poison. So, human will have high chance to eat poison fishes.

Otherwise, such as pollutin case, it can infuence inflation or deflation. Consequently, the reason indicates supply and demand theory. If air pollution is serious, then we will consider health issue, global cars demand number may be influenced to reduce, when global cars number demand will reduce, global car prices and supply number will need to change to fall down in order to attract or persuade global car consumers choose to make car purchase decision.

Hence, global car manufacture number and car price will be influenced to reduce, due to global air pollution issue. Consequently, deflation will occur because when the country citizen usually does not spend much extra saving money to buy car expensive goods. Money value will be low. Otherwise, if global cair pollution is not serious, human considers to buy cars to enjoy driving leisure lives. So, global car demand is influenced to increase , also global car price will also influenced to increase.

Consequently, gobal human will choose to buy cars to drive. Due to we accept to spend extra saving to buy expensive car goods. Car sale price and supply may be influenced to rise up. Money value is influenced to reduce. Inflation may be influenced, due to global car consumers number increases, we would not have extra money to spend easily. Car expensive goods expenditure influences our spending habit to avoid to make car purchase decision more easily. So, human intellectual or foolish activities may bring inflation or deflation consequency in possible indirectly in macro economic view.

On conclusion, above pollution case explain that how and why human intellectual or foolish economic behaviors may bring inflation or deflation consequency as wll as economic growth or recession consequency as well as any goods demand and supply increasing or decreasing consequency. It implies that human behavior may have indirect relationship to influence any goods demand and supply number to either increase or decrease result as well as any goods price will be influenced to increase or decrease in micro and macro economic view.

The relationship between social change and human behavior

Why does economic changes may influence human individual behavioral change? I shall attempt to indicate shopping behavior and staying at home behavior to explain their case and effect relationsip as below:

Human behavior can be influenced by economic change or economic change can be influenced by human behavior? Why does recession may influence consumers reduce shopping desire? In social recession suitation, it is possible that many people lose jobs suddenly, due to businessmen lose many customers. They need to make decision to reduce employees number in order to continue to keep businesses. Consequently, many firms ( organizations) their employees may lose jobs. When they have much time, due to lose jobs, they will feel to avoid to spend too much time and money to go to shopping often. Many losing jobs people, they will often stay at homes. So, they will reduce time to go to shopping, then non essential products won't their preferable choice purchase products. Hence, recession will change many losing jobs people their shopping or consumption desires to avoid to buy non essential products often . Usually when economic boom, many people have jobs to do because consumers number must increase when many people have jobs to do. Then, many people can accept to spend money to buy non essential products often. Many people feel spend time to go to shopping can satisfy their purchase of any kinds of new products useful psychology or desire. So, recession is one good example to explain it can influence many people do not like often to leave homes to go to shopping easily. Many people like to stay at homes, becaue they feel worry about spending too much shopping time when they leave homes. Their staying home time is one good negative shopping behavior example. So,

economic change may influence human individual behavior changes , they have direct cause and efect relationship in behavioral economic view.
May human behavior influence economic change? Is it possible that human behavior may bring the country social economic change in macro economic or micro behavioral economic view ? I shall indicate publishing industry example. Do you feel that if there are many students feel learning is very important when they read many books or many of students feel interesting to read or they have reading new books in habit, then it is possible that the country will have many students like to spend time to go to any book shops to choose the books, they feel that they can help they learn new knowledge. Then the country will increase students number, they often spend time to visit any one book shop every week. Their visiting book shops behavior which may become their habits. So, the country will increase students number, they often spend time to visit book shops. Also, it implies that visiting book shops behaviors may be their behavioral habits.
So, when the country has many students often spend time to visit book shops , their visiting book shops behaviors may help any one book shop to raise books sale chance. So, the country's student individual often visiting book shop behaviors, their habitual visiting book shops behaviors must may assist help any one book shop to increase books sale number absolutely.
Consequently, any one book shop , its books sale bumber must be influenced to increase to increase because the country will have many students like or feel need visit book shops habit in order to choose any suitable books to buy to read at home in order to raise themselves learning effort. When the country has many bok shops often have many students visit their book shops, then their books sale number may be influenced to increase. It explain why student individual visiting book shop behavior may help any one book shop sale number increases also.

How human productive behavior may influence economic development

May any country which citizen behavior assist themselves country development? It is one cause and effect economic question. I mean that if the country itself citicen can not concentrate mind or energy to choose to do one kind of industry in order to let themselves country can bring the most benefit, then whether the counry itself economy can bring the most serious economic benefit. I shall attempt to indicate these countries themselves indistry choice to explain whether these countries themselves citizen productive behavior may help themselves countries to achieve the largest economic benefits. I shall indicate as below:

New Zealand farmer individual wine productive behavior

For New Zealand country example, this country concerns itself effort is foucs on farming agricultural aspect. So, this country has many farmers concentrate on farming agricultural aspect. May New Zealanders choose to spend time to produce different kinds of wines, e.g. wine or red grape wine is for the people are eating meat, or they are eating dinner.

When these New Zealanders their behaviors choose to do farming or agriculture to grow and produce different kinds of taste of white or red grape wine drinking products job. Themselves grape agriculture behavior will influence these New Zealanders themselves, they can learn how to improve different kinds of grape wine drinking products in order to achieve every kinds of white or read grape wines taste improving aim during their white or red grape producing process.

Why can New Zealander every individual white or read grape wine producers improve their white or read grape wine taste more easily? In behavioral economic view, it can explain that why any one New Zealander white or read grape wine producer can be encouraged or excited or persuaded to concentrate nervous and energy and effort to learn how to improve their white or red grape wine products easily.

In fact, New Zealand is one agricultural food export country. It has good natural environment resource , e.g. land, seed to provide any one farmer to produce themselves any kinds of agricultrual food products, e.g. fruit, or wine food products. Because New Zealanders know themselves country has enough natural resource . So, in common, many New Zealanders choose to attempt to do farming agricultural jobs in order to export themselves any kinds of fruit or meat or wine products to overseas or sell to domestic in order to earn profit.

So, when these New Zealand farmers number has been increasing every year. This country farmers will feel themsleves competition between this New Zealand farmers themselves are serious due to they may feel New Zealanders choose to do agriculture businesses in order to export themselves different kinds of farming food to overseas or sell to local to earn profit.

Hence, when many New Zealand farmers feel that farmers number has been increasing every year. They will feel themselves competition is serious. They must need to spend much time and nervous and effort to research what method is the best how to produce the best taste of white or red grape wine products in order to let local or overseas wine buyers to choose to buy

his/her producing white or read grpae products to drink.
Hence, in competition psychological view, may influence many New Zealand white or reaad wine producers had been beginning to change their learning behavior on researching what method is the best in order to produce the best quality of taste red or white wine products to sell in order to attract overseas or local white or read grape wine drinkers to choose to buy his/her wine products. Their behavior will focus on learning how to raising or improving white or read grape wine taste method more than only focus on producing a large number white or red grape wine products. They believe wine quality is more important to compare wine producing number. So, New Zealand wine producers themselves wine producers behaviors have been changing on concentrating on researching wine quality method aspect more then wine producing number aspect in behavioral economic view.

America high technological productive behavior

For America example, US is one high technological country, it owns many high technological knowledge talent inventors, e.g. computer science inventors. Hence, US must attract many diferent countries owning high technological computer inventors choose to go to US to develop their computer science profession career. Also, it seems that when many computer science inventors or professions choose to go to US to develop themselves computer science new career. In behavioral economic view, due to their leaving themselves countries choice, which may bring influence themselve country job behaviors need to be changed. They must need to adapt US new live. Because they will forgive their past computer science job. These computer science professionals need to spend time to adapt US new lives. They " past computer science job behaviors" will need to be changed to their new US any computer employer's new computer science job model.

Because their traditional computer science jobs needed to be forgot in their themselves countries. They will feel their old computer science job knowledge and behavior needed to change in order to let their US any one new of computer company employer feels satisfactory to accept their new working behavior in any one US computer organization.

So, on the other hand, many US computer company employer will feel that they must need time to accept any one new overseas computer science professions their working behaviors, their working attitude daily, because these foreign comouter science professional, their past computer working behaviors and working attitude must be different to US domestic computer science professions.

In behavioral economic view, these overseas computer science professions, their working behaviors and attitude must be needed to change in order to adapt any one US new computer company itself domestic or local computer science professional stafs themselves daily working behaviors and attitude because these overseas and local computer science professionals must need to team work together.

In behavioral economic view, it is only one way that foreign computer science professionals must need to change themselves past country traditiona daily working behaviors and attitude in order to cooperate with these US local computer science professionals in teams more easily.

Consequently, if these foreign compute science professionals can change their past working behaviors and attitude to let any one US local computer science professional feels to cooperate with them easily in short time. Then, the US computer company itself whole computer professional teams themselves efficiencies will be influenced to raised or improved by the changing past working attitude and working behaviors of these foreign computer science professionals. So, in behavioral economic view, only if US any one computer company hopes itself computer teams themselves efficiency can be raised or improved when it decides to employ foreign computer science professionals and US domestic computer science professionals. They need to work in teams together. They must need to let these foreign computer science professionals to know how to change their working behaviors and attitude to let their domestic computer science professionals feel easy to work together. Then, the US computer company itself whole team efficiency must be rasied or improved easily in short time.

- China share market investing behavior

For China share market example, economic development depends on financial market. Because if many Chinese have interest to invest to carry on shares buying and selling activities in orde to learn how to earn shares interest and share profit when the China shareholder can make decision to sell himself/herself shares in the the high price, then he/she can earn money when he/she can sell the China company's shares in the high sale share price position.

If China has many Chinese like to spend time to carry on investing shares activities. Themselves shares buying and selling behaviors will influence China has many companies can increase fund from many Chinese shareholders in order to have enough money to expand or develop themselves businesses in China in long term.

Consequently, when China can have many Chinese like to attempt to carry on buying and selling shares investing behaviors in China share market. Themselves buying and selling shares behaviors can help many Chinese companies have effort to increase enough money or capital in order to continue to do their businesses in long term absolutely. So, it explains why when many Chinese become shareholders , they can assist China will have many companies continue to develop their businesses if many Chinese like to carry on shares buying and selling investing behaviors in long time in China financial investment market nowadays in behavioral economic view.

Why has any individual country have many people invest share behavior which can influence the country's macro consumption desire?

I shall apply shares market buying and selling investment behavior to explaiin why shares investment behavior which may impact the country's overal consumption desire as below:

In behavioral economic view, I assume that when the coutry has many people have interest to attempt to carry on shares buying and selling investment behavior, then their frequent shares buying and selling behaviors which may bring negactive consumption desire or shopping desire of these shares investors their consumer behavior.

The reason is simple, when the country has many share buyers number suddenly been increasing rapidly. Consequently, these large group share investors must need to spend much time to research any kinds of company shares variations, whether when their share prices will rise up of fall down in order to achieve buying the company's shares in the lowest price and selling the company's shares in the highest price level in order to earn profit.

Basic on this reason, they must need to spend much extra time to research share prices changing behavior every day, e.g. one working person will wait to leave his/her job, after he/she can spend time to gather data to research the day's share price changing behavior after dinner. So, the working person's right time may be his/her share price market research behavior. Before he/she may spend his/her night time to go to shopping after dinner, but nowadays, he/she will fogive to do his/her shopping behavior before dinner or after dinner at hight sometime. He/she will make decision to spend much night time to turn on computer to click on share market website to research his/her share purchase choice to investigate whether his/her share price whether it rises up or falls down at the moment in order to make his/her share buying or selling decision at ever night time.

I mean the when the country has many people are share investors, their shares investment behavioral spenging time which will influence many shops lose customers at might often because the country will have many people feel need to spend night time to turn on computer or watch television to investigate share price variation. So, the country will have many people / share investors choose to stay at home in order to carry on share price variation investigation behavior, they need to listen share market update news from radios or watch the share market update news from computer or TV at home every night. Consequenly, they must reduce times to leave themselves homes at night. So, their shopping behavior also will be reduced. Because these share investors feel need to spend time to investigate share price variation news at homes which can bring economic benefits ( high opportunity benefits) when they choose to forgive to leave homes to go to shopping times ( opportunity cost) every night.

On conclusion, it seems that when the country has many people are share investors, then their share price investigating behavior may bring negative shopping emotion at night. Consequently, the country's any one shop may lose many customers from this share investor consumer group in behavioral economic view. Hence, when the country's share investors number had been increasing rapidly, it will influence any shops lose many customers from this share investing customer group at night frequenly in short time, even long time in behavioral economic view, because their shopping desires or shopping emotion will be brought negative feeling when they make decisions to spend much time to listen radios or watch TV or computers share price update nes at night. Hence, share market will bring negative impact to influence consumer shopping desire or negative shopping emotion in behavioral economic view.

Can technology influence human shopping behavioral change?

Nowadays, technological development has reached mature stage, whether technological mature stage may bring positive or negative shopping emotion influence to global consumers. I shall aplly internet inventin or ecommerce shopping channel tool to explain whether internet technology can bring postive or negative influence to global consumer behavior in behavioral economic view.

Internet is a good technological tool, it brings e-commerce business chance.

In fact, commonly, global has have many businessmen choose to use internet channel to carry on their products transactions between global online-buyers and their electronic websites. So, global many shoppers had begun to feel online shopping is more convenient to compare visiting shops shopping. Their shopping behaviors have been changed from internet technological tool. Global has many shoppers choose to buy any products from any overseas or local businessmen their web stores. They only need to spend time to find any businessmen their webstores to choose the most suitable products to pay visa to buy from their webstores. at homes. So, in general, global had have may shoppers had changed their shopping behaviors from visiting shops to visiting webstores at homes often.

So, it seems that internet technological tool had influenced global many shops disappear, but internet webstores will be replaced their actual shops on streets. Some of businessmen either they choose webstores to replace shops or choose websotes and shops both or still keep shops only. Hence, internet tool influences global businessmen have three kinds of products sale channels to let globa local and overseas consumers to choose how to buy their products.

However, in fact, many of global shoppers, youngers and olders had begun to accept to buy any products from webstores. They feel to spend time to leave homes to visit shops , their shopping behaviors will be wasted time to not essential part to their daily lives. Hence, since internet technological invention, it had changed many consumers their traditional visiting shops shopping habit to change to buying products from webstores channel.

However, on the one hand, internet creates webstores ecommerce shopping channel to let global many consumers do not need to leave homes to go to shopping. It brings negative visiting shops shopping emotion to global general consumers nowadays. But on the other hand, it also brings positive visiting internet webstores shopping emotion to global general consumer nowadays. So, it seems that global many consumers feel that they often do not need to spend much time to go out shopping. Many global consumers feel convenient and enjoy to choose any products to buy from different internet webstores, when the online buyer chooses the most suitable product, he she only needs to pay visa card to buy the product from the online seller's webstore conveniently at home.

Hence, online shopping can bring economic benefit to online buyers, e.g. avoiding walking time or spending transport fare to visit the shop to go to shopping, shortening or reducing shopping time to do another important matter.

On conclusion, global many consumers began feel online shopping can bring more economic benefits on shortening shopping time, avoiding transport fare spending aspect. So, online shopping will be popular shopping behavior for future long time. It may encourage global many shoppers can make rapid shopping decision in short time in order to carry on any products buying transaction to global any one online shopper in short time easily in behavioral economic view. So, global many businessmen had begun to build themselves one attraction webstore in order to persuade different countries consumers to choose to click themselves webstores from internet channel to buy any kinds of products in short time easily.

So, internet technology had changed consumers traditional shopping behaviors to build positive online shopping emotion as well as raise online sellers' any products sale chance easily in behavioral economic view.

Why and how human behavior may influence the country's economic growth or recession?

When one country has many people choose to do the same matter for one period, whether their behavior may influence the country's pvera; economic growth or recession . I shall attempt to indicate cases toexplain their relationship as below:

For flowing rubblish behavioral case example, do you feel that when the country has many people often flow rubblish on the streets, instead of their flowing rubblish behavior may bring streets dirty? But, their flowing rubblish behavior may explain that this country has people may have enough money to buy food to ear, or enough cloths to wear, enough bottles of water to drink, even they may have enough money to buy new television, radio, refrigeraters , washing machines, desktops or laptops electronic home products from old to new to use in order to satisfy their living needs. So, when they flow old electronic home products, their flowing old home electronic products behaviors may seem that they have enough money to buy other new home electronic products to replace old home electronic products to use at homes.

However, it seems thaat this country ought have many people have jobs to do. So, many of them, they can easy to make purchase decison to flow any old home electronic products and buy any new home electronic products to use . Because this country has many people have jobs to do. So, they can often not use old home electonic products to become rubblishs to flow on streets after they had bought any kinds of new home electronic homes.

In fact, it also implies that this country's economy grows rapidly. So, many businesses can glow up rapdly. When they expanded their businesses, they must need to increase employees number in order to let they help themselves to raise productivity or serve their clients absolutely. So, when the country has many businesses can grow up, it seems that its economy must be better or it is improved to compare past. Due to many different kinds of home electronic products had been often bought to use by this country people in this period. So, this country's any streets can be observed that expensive electronic home products were flowed on streets anywhere. then, this country will have many electronic home products sellers can sell their home electronic products very easily. When this country has many people can find any kinds of jobs to do easily. So, due to unemploymen rate had been decreasing.

In behavioral economic view, as this many electronic home products rubblish country case, we can observe this country may have many people have jobs to do. So, consumption number has been increased long time. So, cheap food, or expensive home electronic products may be rubblish on any streets. This country's people , their flowing rubblish behaviors may be explained that many of people have enough jobs to do, so they have ability to buy any good taste food to eat or buy any kinds of expensive electronic home products to use. So, this country's economy may be improved for this long period. So, in behavioral economic view, when this country can have many electronic home products rubblishs are flowed on anywherer in streets frequently. It seems that this country will have many people have jobs to do, so it causes they often change old home electronic products or replaced them easily, when they have enough income to spend to buy any kinds of new home electronic products to use at homes easily. Moreover, their flowing old electronic home products behaviors also indicate that this country has many people their salaries may be increased in possible from their emplyers. When this country can have many different kinds of home electornic products are sold. It means that this country's electronic home products needs or demand had been increasing, due to many people have jobs to do and income increases to excite their living of needs also improve. Consequently, this country may seem have better economic improvement. We can observe from this country's electronic home products rubblish increasing income in theis period.

On conclusion, this country ought experience economic growth at this period. So, " flowing expensive electronic home rubblish increasing number

" may seem that this country's economic growth is rapidly in this period, due to many people have jobs to do as well as salaries increase in this period.

Technology how impacts human behavior changing?
Technology how influences human behavior to bring changing? For example, online share purchase and sale transaction from smart phone brings share investor can do share buying or selling transation in any where and any time conveniently, non manual driving auto vehicle, bring car owner feels comfortable and spends free time to do other matter, e.g. reading, listening mucis in himself or herself car freely. electrical energy vehicle can help car owner to reduce air polluton and it can brings the drivers do not feel drive long time in any journeys in order to avoid air pollution for environmental protection responsible car drivers in our societies. Thus, they will drive long time in any journeys when they can drive electronic energy cars to replace oil energy cars.
However, online technology can also bring consumers can choose to stay at homes to buy any things from seller individual online webstore conveniently. Such as online technology can bring shoppers do not need to spend much time to visit shops to buy any things. They can choose any kinds of products from any online sellers individual online webstores conveniently at homes. Online technology excite busy consumers can make purchase decision easily as well as it can help online sellers sell any kinds of products from internet easily.
In behavioral economic view, technology can change human behavior to be improved, it can let human feels comfortable, more free time ro use, rapid making any decisions, such as apply smart phones to make share purchase or sale transaction decision, online shopping decision, even travelling any where decision in short time, when the traveller finds the most cheap hotel accommodation room price and air ticket price frm any travel agent online tourism webstore, then the potential travel customer can follow the online hotel accommodation price and air ticket price data to make decision when to buy the air ticket from the airline travel agent or make decision when to prebook which hotel accommodation room to go to the country to travel from online travel agent tourism webstores. So, technology can encourage global any country travelers to make anywhere to trvel rapidly. If the traveler can find the country's general hotel rooms and airline tickets prices had been decreasing more sightly. The traveler may make travel decision to choose the country to travel in short time, then he/she can

prebook the country;s any hotel room and airline ticket to pay by visa fraom the country's any hotel and airline travel agent webstores., before one week, even one month or more easily. Hence, online technology can also encourage traveler individual frequent travel times to be increased, due to global travelers can find any hotel rooms and airline tickets prices from internet conveniently at homes. They do not need to spend time to visit any airline travel agent to enquire travel choice country's hotel rooms prices and airline ticket prices. They can compare global travel of countries choices ' all hotels rooms and airline agents air tickets prices to make prebook airline seat and hotel room decision before one week, one month even six months early.

On conclusion, online technology can encourage global travelers can make travelling any where and when traveling time desicions easily. It can excite tourism industry develops in long time. Also, such as electricity cars invention can encourage environment protection car owners do car purchase decision easily, because they can choose to drive electronic energy cars to replace oil energy cars in order to avoid air pollution occurs easily. So, electronic cars can increase electronic car purchasrs number, due to many of environmental protection attitude of car owners can choose to drive electricity cars to bring air cleans, even non -manual driving cars can encourage lazy driving and free time driving car owners to choose to buy non-manual ( artificial intelligent) cars to drive , because they can spend much free time to read, listen music or do any matters in themselves cars, they do not need to drive cars, robotic (AI) auto driving machine is such one non-manual driver to help them to drive themselves cars confidently. So, non-manual driving cars can attract lazy and enjoying free time driving car owners to choose to buy to replace traditional manual cars to drive easily. Moreover, online share transaction can help any share investors to make share buying and selling decision in short time easily. When they can apply smart phones technological tool to carry on share buying and selling activities easily. They can observe any share rising or falling price suitation from smart phones in any where any any time easily. So, smart phone technology can help global any shareholders to make share purchase and sale transaction easily. So, technology can encourage human makes decision in short time rapidly.

How and why employees behaviors may influence economy development?

In behavioral economy view,I believe the country's any organizational employees behavior may bring indirect relationship to influence the country's long term economic development. I shall indicate past manufacture industry social development period to explain their relationship. For many countries' past business activities had belonged to manufacturing industry, such as US, UK past before 1980 year, it focused on steel manufacturing and steel manufacturing related machine products. So, US, Uk developed countries manufacturing industries may be past main country's economic income sources. I assume US , UK past had one million number different kinds of industries. They ought had about seven houndred thousand number organizational businesses were belonged to manufactured industry. They may include:
Steel manufacturing and steel related machine manufacturing, e.g. vehicle manufacturing, home appliances, e.g. washing machine, television, radio, refrigerate cooler, heater, air condition etc. different kinds of different kinds of steel -related manufacturing machine, they were manufactured from US, UK steel machine manufacturers. So, US, Uk the other three hundred thousand number industry may be general service industry, e.g. hotel service, restaurent, cinema, public transport service, tourism lesiure , wine bar, supermarket etc. different kinds of non-manufacturing industries business organizations were operated in UK, US past before 1980 year.
So, in UK, US developed countries industry development history, they ought have high percentage of businesses belonged to steel related manufacturing machine and steel products. Also, in the past before 1980 year, US, Uk business employers , they employed many workers are manufacturing workers. They needed to spend long time to work in factories. They were skillful workers, and they are trained to manufacturing cars, washing machine, television, heater, etc. even steel itself different kinds of steel related products to prepare to deliver to their shops to sell to US, Uk local or overseas clients.
So, I believe that past UK, US ought employ many employees, they belonged to skillful manufacturing workers, manufacture increasing steel machine or steel related machine number of products rapidly daily. So, if UK, US had had many of these manufacturing factories owned high skillful workers, then their manufacturing steel-related machine or steel both kinds of products number must be influenced to raise rapidly. Consequently, their steel machine manufacturing products would been exported to overseas or would been sold to local both markets , they may be influenced to raise

sale number. They ( these manufacturing workers) needed to be trained to know how to manufactur these different kinds of machine products in the efficient teams and they ought to be trained to raise their efficiencies in order to shorten time to manufacturing many kinds of steel related manufacturing machine or steel itself products rapidly. So , if their efficiencies and manufacturing performance was improved, these US, UK any one manufacturing worker and their teams ought achieve raising productivities significantly.

Hence, when past UK, US manufacturing industry development period, if these two countries' any manufacturing factories could have many manufacturing workers could be trained to be skillful and proficient manufacturing workers. Then, in past every day to these factories workers, they ought help their steel or steel related manufacturing employers to raise any kinds of machine or steel products number in every team. So, when past in the manufacturing industry development, US, UK could have many factories' manufacturing workers themselves steel or steel related machine products manufacturing skill could be trained to to improve to any kinds of these machine or steel manufacuring products quality as well as their products number could be influenced to raise by themselves skillful improvement significantly every day.

Then, what would be influenced to occur to past UK, US manufacturing industry period? In behavioral economic view, when these two manufacturing industry developed countries, such as UK, US , if they had many factories workers can be trained to improve their skill in order to achieve any kinds of steel or steel-related machine products quality could be improved as well as products manufacturing number could be also increased absolutely.

In consequence, past UK and US both countries ought increase themselves any kinds of steel and steel related machine products number to be supplied to themselves local shops to let local clients to choose any one kind of machine manufacturing products to buy easily as well as they could also export to supply overseas any countries to buy their different kinds of steel or steel related machine products to let overseas steel or steel related manufacturing machine product buyers, they can have many of these different kinds of these steel or steel-related different kinds of manufacturing machine from UK and UK these both countries easily to compare other countries.

On conclusion, I believe that past US, and UK macro manufacturing industry income GDP would increase significantly. So, they would have good economic growth performance because when many of these manufacturing workers themselves manufacturing effort could be improved. So, it explained when employees manufacturing abilities can influence economic growth indirectly.

Robots invention whether they can help organizations to raise efficiencies or inefficiencies?

In behavioral economic view, in any organizations, when the organization hopes its worker teams can raise efficiencies , the organization may choose to increase more workers number and/or it can provide training to improve these workets themselves skills in order to raise their efficiencies. For one warehouse example, when the warehouse increases many goods , they are needed to delivered these goods from the shelves to the delivering destination locations. If this warehouse supervisors feel these workers themselves goods delivery speeds are slow, which is possible due to this warehouse's workers number is not enough. So, this warehouse supervisor ought increase workers number in order to increase their goods delivery speed in order to deliver goods from the shelves to every indicated goods delivery destination in order to let any one lorry driver can transport the right kinds of goods and ensure the accurate goods number to transport to any one client home rapidly.

However, if this warehouse supervisor planed to buy several warehouse goods delivery robots to assist these warehouse workers to find the right kinds of goods from shelves and then deliver to the right destination location in the warehouse. So, these warehouse orkers can concentrate on counting the accurate goods number and ensuring the right kinds of goods in order to prepare to let lorry drivers to transport these goods to these goods of buyers themselvers homes rapidly. Consequently, in the first step, robots can concentrate on finding th right goods from shelves and delivers them to the right goods transportation of location destination. Then, in the second step, these warehouse workers can concentrate on counting the accurate goods number and ensuring the right kinds of goods in order to prepare to put them to the lorry. Consequently, when warehouse robots and warehouse workers can cooperate to work together, the most important, robots, can deal on finding the right kinds of goods and deal on delivering the accurate number of goods of job duty as well as these warehouse workers can only concentrte on counting the right kinds of goods number in order to avoid it has none any mistake of wrong kinds of goods and inaccurate goods of delivery number to be transported to the lorry and to deliver to any one buyer's home.

So, it seems that warehouse robots ought help any one warehouse worker to raise himself efficiency and avoid goods delivery of mistake occurrence easily as well as their help to warehouse workers that can let any one goods buyer feels their goods can be delivered to their homes rapidly. Moreover, warehouse robots can also help these warehouse workers to raise efficiencies because warehouse robots can help them to shorten goods delivery time between any one shelf and any one goods delivery destination of location in the warehuse because robots may help them to find the right kinds of goods from the right shelf in the short time. So, any one worker does not need to spend long time to seek anywhere is the right shelf location for the kind of goods when the kind of goods are needed to deliver to the buyer's home from lorry. Warehouse robots can help them to do this aspect of " finding the goods from the right shelf in short time job duty". So, any one warehouse worker only needed tospend less time to do the counting of any right kind of goods number and ensuring the right kind of goods job duty. Consequently, this warehouse 's any one worker, his any one kind of goods delivery time may be reduced, because robots‘ assistance and they may have more confidence to avoid mistake to deliver the wrong number of goods and/or the wrong kind of goods to any one goods buyer's home.

On conclusion, it seems that warehouse robots ought may help any one warehouse worker to raise efficiency for any one team in the warehouse as well as the warehouse any one supervisor does not need to spend much time to observe any one worker individual performance for " goods delivery job duty aspect" because their goods delivery job duty that had been replaced to do by these several warehouse robots. Robots can achieve the more accurate of right kinds of goods and the right number of goods delviery job performance to compare any one of human warehouse worker themselves right kinds of goods of delivery and right number of goods of delivery job performance. So, when robots can participate to cooperate with this warehouse's any one worker to do their goods of delivery job duty in this warehouse every day. Then, robots can raies any one of supervisor individual confidence in order to let they do not need to spend time to observe any one of worker individual whose goods of delivery job performane. They can concentrate on supervising any one worker whose goods transport to lorry in the final step in order to avoid to deliver wrong goods number and / or wrong kind of goods to any one goods buyer's home every day. Consequently, this warehouse's overall teams of their delviery of goods performance many be improved by robotss' participatin to goods of delivery task as well as this warehouse's oveall teams themselves efficiencies may be influenced to raise by robots' goods of delivery task participation.

Why social behavior may influence organizational strategy needs to be changed ?

Why any organizations need to know whether nowadays social behaivor how has been changing in order to implement the kind of the most right strategy to achieve the profit aim pursue in possible. I shall indicate nowadays ecommerce or online, customer shopping behavior to explain above question concerns they ought have close relationship between social behavior and organizational strategic choice or organizational behavioral changing need.

On nowadays ecommerce business, or online shopping model, this kind of shopping model in global many young and old age consumers like to apply internet tool to choose any country sellers website stores in order to stay at home to buy any kinds of products from themselves webstores in global societies.

In fact, online shopping model had been popular for long time above to twenty years. Most of global sellers will make decision to design themselves webstores in order to attract global many online buyers to choose to buy their products from themselves webstores. So, it seems that social consumers purchase behaviors had been changed to online shopping from internet invention.

Hence, social consumers purchase behavioral changes may influence any organizations' strategies need to be changed from visiting shops purchase strategy model to online purchase strategy model, if the seller still concentrate on concentrate on considerate how to design itelf , but neglects to considerate how to design itself webstore, e.g. how to design attract product photos to put on itself webstore, how to arrange sale price information location to be putted on webstore and visa card payment location on itself webstore in order to let any one online buyer can feel very easier to buy itself any kinds of products from itself webstore. Then, its potential online buyers will be influenced to increase number when they can find this online seller itself any kinds of products photes and every kinds of product sale price information and visa card payment channel

locations easily from itself webstore.
So, it implies that nowadays any one seller ought need to design one webstore to let any one online overseas and domestic consumers can have chance to click itself webstore to choose any one kind of product to buy conveniently when he/she does not hope to leave him/her home to go to shop, because nowadays social shopping behaviors had been influenced to change when internet invention, them it gives another online purchase method to replace visiting shops purchase method to global any one buyer in nowadays societies.
So, if nowadays any one seller still concentrate on how to design itself shop display in order to put any kinds of product on shelf in order to let any one visiting shop customer to find the kind of product to buy, but it neglects to change to choose to pursue another new technological shopping method, such as webstore purchase method in order to implement effective strategy to design the most right webstore as well as in order to attract global overseas and local consumers to find itself webstore easily from website and find its any one kind of product phots and sale price and visa card payment button in order to choose to buy itself any kinds of products in the short time. Consequently I believe that the seller will lose many customers from overseas and local when its other same or similar product sellers choose to design themselves webstores in order to let global any one product buyer can buy themselves any one kind of product when they can pay visa card to buy their products from them webstores conveniently when they stay at home habitly. Then, the seller will lose many global potential customers in long time.
On conclusion, in behavioral economic view, any consumer behavioral social changing, which will influence any in order to avoid customers number loses significantly . In future time, organizations need to make rapid decision in order to implement the most reasonable and the most useful strategy in order to avoid global potential customers number reduces or lose them in long time. So, social behavioral changing environment ought influence any global organizations need to decide how to change themselves strategies in order to avoid customers loses significantly in future time.

How and why human behavior may influence economic growth or recession?
May ourselves daily behaviors influence our global societial continue economic growth or recession? Do they have cause and effect close

relationship between human behaviors and global economic growth or recession? I shall apply behavioral economic theory to analyze and explain whether ourselves daily behaviors and our global societial economic growth or recession which have close cause and effect relationship as below:

Every country itself economic development must depend on any business activities, otherwise, any kinds of business activities must need ourselves business activities or behaviors in order to achieve any business activities as well as achieve the country's overall economic development in macro view. However, any country's overall business activites or behaviors which must depend on any kinds of individual businessmen, themselves employees daily working behavior or activity or performance in order to help them to attract or increase many clients number to acieve " earning profit" aim. So, it seems that any individual business, itself overall every department individual working behavior is one main factor to influence the company's overall business performance.

For agricultural fruit and meat food farming industry example, such as New Zealand is a farming main target industry country. It had had many New Zealanders were daily themselves own farming businesses for many years. Their farming businesses include growing fruit, sheep, cow, pig pork, meat etc. food sale business. If the New Zealand farmer owned a large size farming land, then he will choose either growing fruit or feeding sheeps, pigs, cows to be meat to to transport to New Zealand supermarkets to help them to sell to their farmers meet to New Zealanders in order to earn profit. Thus, if the New Zealand farmer owned large size of farming lands, then he needs to employ many farming employees ( farming workers) to help him to carry on farming business daily tasks, e.g. picking up friuts, feeding pigs, cows, sheeps to eat food daily. These daily farming jobs are very important to influence this New Zealand farmer's meats or fruits sale number whether they can be easy or diffcult to sell in New Zealand supermarkets , if these farming workers can own encough farming knowledge or skill to know how to pick up fruits method and make judgement to know whether it is right time to pick up the kind of fruits from the trees , as well as know how feed this pigs, sheeps, cows to eat food in order to let they are better health. Consequently, their farming behaviors which can let these animals can provide the best taste and enough meat from these animals to let New Zealander to buy to eat from New Zealand any one supermarket. Even these New Zealand farming workers can know whether the kinds of fruits, e.g. oranges, apples, gapes etc. fruits whether they ought be picked up from the trees at the right time. Consequently, they can make judgement to decide to pick up any kinds of the best taste fruits to let any one New Zealander to buy to eat from any one supermarket in New Zealand. Otherwise, if they do not make judegement to know whether the kind of fruit ought not be picked up because they still need longer time to continue grow up to increase fruit size and better taste from the trees in order to let any one fruit buyer can feel better taste when they eat this kind of fruit later. If they can buy this kind of fruit to eat later, then this New Zealand farmer's his fruit buyers can buy the best taste of this kind of fruit to eat from an yone supermarket in New Zealand. Consequently, many New Zealand supermarkets will choose to buy any kinds of fruits from this farmer fruit supplier when they feel this farmer's fruits can provide more better taste fruits to compare other farmers‘ fruits.

Thus, due to New Zealand is one farming main income source country. It's any kinds of fruits and meats need to be export to overseas to sell , instead of local sale. It's GDP percent is very high to whole country 's overall income source. So, any one New Zealand farmer individual and any one farming worker individual working behavior will influence its economy whether it is influenced to grow or recession possible. Moreover, it also seems that farming workers' farming knowledge and skill will influence themselves farming daily activities to achieve the aim of the number of increase or decrease to any kinds of fruits whether they are better taste or the number of increase of decrease to any kinds of meats whether they are better taste to supply to any one New Zealand fruit or meat buyers to eat from any one New Zealand supermarket. So, it implies that any one New Zealand farming worker individual farming behavior may influence any kinds of fruits or any kinds of meat taste because they are transported to any one supermarket to sell in New Zealand.

Consequently, if New Zealans had many farmers can teach god farming knowledge and skill to let their any one farming workers know how to decide judgement to decide when it is right time to pick up any kinds of fruits from trees , or how to grow them on soil in order to let they can grow rapidly. Then, many different kinds of fruits can be provided to let any one New Zealanders can eat the best taste of fruits when their fruits are supplied to any one New Zealand supermarkets. Even, if they knew how to feed foods to pigs, cows, sheeps to eat daily. Then they can be more health and they can provide the best taste of meats to let any one New Zealanders can buy their meats from any one New Zealand supermarkets. Moreover, their fruits and meats can be transported to overseas to let any one country fruits or meats buyers can choose any kinds of New Zealand meats and fruits to buy to eat from themselves countries supermarkets. Then, many overseas fruit and meat buyers will perfer to choose New Zealand any kinds of fruits or meats to buy to compare other countries fruits or meats to buy when they go to any one local supermarkets.

On conclusion, it seems that New Zealand farming workers themselves farming behavior may influence their farming employers any kinds of fruits or meats sale number and income because their farming task behaviors must influence whether their fruits or meats taste are the better taste or worse taste to compare their other local farmers ( the farmer competitors) whose fruits or meats taste. If tthe farmer's any one farming worker can be trained to learn how to know to feed animals skill and when is the most right time to pick up any kinds of fruits from trees or how to grow them on the soil methods. Due to these farming worker individual farming behavior may influence his different finds of fruits and meats sale number to be increase or decrease, so these any one New Zealand farmer must need to depend on any one farming worker whose farming working methods, if their farming working behaviors can be the best to influence any kinds of fruits to grow rapid or any kinds of pigs, cows, sheeps animals grow up rapidly , then their sale number may be increase significantly and their taste can be improved to let any New Zealand or overseas meat or fruit buyer to buy to eat to feel from any one New Zealand or overseas supermarkets, then New Zealand's agriculture industry must be influenced to increase. In the world, any one fruit or meat buyer must choose to buy New Zealand's fruit and meat to eat in prefer to compare other countries' fruits and meats. So, New Zealand's GDP may be influenced to raise from any one New Zealand farming worker individual farming working behaviors.

Reasons why human behavior may influence economic recession or growth?

Can ourselves daily behaviors or activies influence ourselves countries' economic growth or recession? I shall attempt to explain the reasons why they have direct or indirect relationship between human behavior and economy growth or recession as below:

I shall indicate environment pollution case to attempt to explain above question. Our societies had been experiencing servious environment pollution challenge. However, environment pollution , such as air pollution is caused by air planes and vehicles emission by air planes and vehicles emission as well as water pollution is caused by plastic rubblish, or dirty water or oil or gas chemical material, these both kinds of pollution ought may bring economic recession and this both kinds of pollution are caused by human ourselves daily foolish activities.

I believe human behavior and economy and pollution which have cause and effect relationship. I shall analyze this environment pollution case to explain why they have case and effect relationship between human foolish behavior and environment pollution and economic recession as below:

When global societies had many people like to buy cars to drive to bring emission to fresh air on the roads as well as many manufacturing factories will bring emission to pollute fresh air in their manufacturing processes. Factories and cars will bring air pollution , due to factories need to pollute fresh air in order to manufacture many products and car owners need to drive their cars to go to offices or leisure places. Their cars will also bring emisson to pollute fresh air. On consequence, car owners themselves frequent driving behaviors and factory workers themselves frequent manufacturing behaviors may bring environment pollution. Technology or human behavior whether may influence economic growth or recession. Moreover, air planes also brings emission to pollute air when they are flying in sky. Also, when ships bring oil pollution or sea plastic rubblishs bring pollution to global oceans.

In fact, manufactuers and cars owners, such as factories workers manufacturing behaviours ans car owners driving behaviors and pilots driving air planes flying behaviors and ships transport behaviors, which may cause plastic rubblish, oil or gas emission to sky or sea or on the road to cause ocean and air pollution is serious. However, human ourselves need to buy cars to drive to satisfy ourselves driving leisure or enjoyment, travelers need to catch air planes to travel to enjoy leisure needs, factories workers need help factories to manufacture many products to sell to customers to satisfy their using needs. oil exploration needs to find lands to explore new oil lands.

All of these business and leisure activites may bring serious air and water pollution. However, due to serious air and water pollution will bring earth warming challenge , such as some countries temperature will be influences to rise up to 40 degree or higher br earth warming. However, earth warming is caused by air and ocean pollution. Pollution must be caused by human ourselves, driving cars leisure and factories manufacturing business activities. Hence, if human decided to continue to do these foolish behaviors, we only pursue to manufacture different kinds of industrial products or drive cars to enjoy leisure aims, but we also neglect ourselves behaviors may bring environment pollution. Then, earth warming or earth temperature will be influenced to rise up absolutely in long term. Moreover, if our future earth will be influenced to bring serious high temperature effect by human ourselves these foolish behaviors.

On consequencey, warth warming will bring serious economic losses in possible because when ourselves earth temperature had been influenced to rise up to 40 degree or high. Ourselves health will be caused poor, due to we will feel difficult breath, we must need often tried and hard to work, due to our nervous and health will be influenced to poor by pollution and earth warming effect. Also, we need to pay more money to see doctors when we had long life. Then, our societies will lose may strong labors to help manufacturers to work, e.g. factories will reduce workers number to help manufacturers to produce more different kinds of products, due to workers health is general poor. Due to lacking enough workers to manufacture products, our societies will begin to reduce enough supply number of products to sell to global consumers to satisfy their use needs.

On conclusion, in behaviroal economic view, our societies will lose many labors due to their bodies are not health by air and water pollution. Global economic and business activities will be influenced to worse by global workers reducing number reason. So, economic recession will begin to occur in possible when pollution reaches the serious level.

How employee behavior influences organizational development?

Can any organizational department employee individual behavior may help the organization to bring long term development? When one employee individual behavior, manager won't feel whose task behavior may help organizational development, but when the department has many teams cooperate to work together , all of these team employees whose task behaviors may help their organization to bring long term development.

I shall explain how any why when the organization has many departments, as well as when every team memmber individual behavior may help whole organization to bring long term development in possible as below:

Every organization must need efficient department to cooperate to work together. They may include human resource, finance, logistic, facility management, sales, marketing , operateional , warehouse , factory manufacture , research and development, purchase, customer service etc. different kinds of departments to cooperate to work together. So, any one employee individual behavior, include manager, leader, supervisor, worker, salesperson, manufacture worker, adminisration staff, factory or logistic worker etc. themselves task behavior whether his/her performance is worse or better , whose task behavior ought bring long term good or bad influence to cause the organization's whose efficiency, or performance , whether it can be influenced to improve significantly. For car factory manufacture workers department example, it exmploys 100 car manufacturing workers. They need to manufacture at least 50 cars in order to bring enough car manufacture number to supply to global car buyers to choose to buy ( satisfaction to car buyers their driving leisure activity needs). However, if this car manufacture firm employs many low skilful car manufacture workers, their inefficient car skill may bring cars manufacture number reduces, they can not achieve to reach the at least 50 cars manufacture number, if these 100 car manufacture workers. They have half number of workers, they only manufacture 30 to 40 cars number at least daily. So, it seems that this car manufacture firm will have half car manufacture workers bring the low cars manufacture number to compare the another half cars manufacture workers, when this proficient car manufacture workers may manufacture at least 60 or more cars manufacture number daily. So, it explains that this inefficient car manufacture workers will not help this car manufacture company to manufacture enough cars number in order to supply to global car market to sell to satisfy global car buyers needs, when car buyers demand number is more thn car manufacture supply number in supply and demand view. Hence, in long term, if this car manufacture company can not employ new proficient car manufacture workers to replace those inefficient or low skillful car workers. Consequently, its car manufacture number must be influenced to reduce and it can not satisfy global car buyers driving leisure needs.

However, if this car manufacture firm also has shop to sell itself any kinds of cars, instead of manufacturing cars product. So, it needs have both main departments to help it to earn profit. The first step, it needs have proficient car manufacture workers to help it to manufacture at least 50 cars from every car worker in order to have enough cars number to be provided to global car sellers to help it to sell to global car customers. Second step, if it decided to attempt to sell itself cars. Then, it needs to set up car shops in global to different countries in order to let global car buyers may visit its global any one car shop to enquire any one car etc. salesperson about any car quality, speed, gas useful, price, safety, etc. information questions and they can attempt to sit in any one car to feel whether which car can let them to feel more comfortable to make final car purchase decision in any one shop. So, if this car company can provide good sale speaking skillful training to any one car salesperson to let his/her to know whether how to explain every kind of car function and feature, manufacture method etc. questions, then I believe that they can influence any one car buyer to makecar purchase choice decision more easily. So, it this car manufacturer hopes it may attempt to earn profit from different countries car sellers and car buyers both. It ought also provide training course to all general car salespeople to be proficient owning sale speaking skillful professional skill in order to prepare having more confidence to persuade any one car customer to make car purchase choice from any one car salesperson more easily to compare global other car sellers.

Hence, if this car manufacturer could build both car manufacturing team and car sale team more proficient. However, if this car manufacturer hopes to develop itself car manufacture busness to expend to car sale business both in success. It must need to spend long term to provide training courses to general car manufacture workers and general car salespeople both to be proficient car skillful manufacture workers and proficient car skillful salespeople in order to help they can manufacture enough car numbers and help they can persuade may car customers can make car purchase decision in short time when they visit its any one car shop.

However, this car manufacture company explains why every car manufacture worker whose manufacturing behavior and every car salesperson sale persuading speaking ability may help this car manufacture company to expand from its car manufacture market to car sale market development in sussess in possible. So, this car manufacture firm must need these two kinds of essential human resource elements in order to achieve its cars sale number and cars manufacture number increasing aim. They may include proficient car manufacture workers and proficient car salespeople both human resource elements. These both human resource daily task behavior may influence its long term task efficient performance in order to expand itself car sale business in success from itself car manufacture business easily. If it hopes to expand its car manufacture business to car sale business in success. It must need to provide training to these two departments general staffs to be proficient staffs in order to supply enough cars number to its global car shops to let global car buyers

can choose its any kinds of cars to buy in any time.
Morevoer, if this car manufacture company can have good skillful of car research and development department , it aims to research and innovate any new technological cars invention in order to improve its any traditional old kinds of cars to be innovative new kinds of cars from every year. Consequently, its any new innovative cars ought attract global any one car buyer to make car purchase choice final decision more easily, because its any kinds of manufacturng cars can be innovated rapidly to compare its any one car manufacturing competitors, when its nay kinds of cars can be shorten time to innovate within three months, but its any one car manufacturing competitors need to spend more than three months, even one year to innovate themselves traditional old cars products in long term.
Hence, its car staffs research and development department staffs must need own good car product design ability, proficient car engineering knowledge , even car invention knowledge in order to innovate its any one kind of car product in short time and introduce to let its global car proficient car buyers feel surprise to its any one kind of innovative car products to compare its any one car manufacturer.Hence, these four departments: car manufacture, car sale and car research and development anr car training departments must need concentrate resource to provide enough training to any one staffs in order to achieve the best performance.
On conclusion, all these departments staffs their performance can influence car manufacture aim to chance to car manufacture and sale aim more significantly. it explains why some main department staffs whole behaviors may influence any organizational performance significantly.

www.ingramcontent.com/pod-product-compliance
Ingram Content Group UK Ltd.
Pitfield, Milton Keynes, MK11 3LW, UK
UKHW021912190726
13853UKWH00002B/629